Framing Technology

Framing Technology

Society, choice and change

Edited by Lelia Green and Roger Guinery

ALLEN & UNWIN

First published in 1994
Allen & Unwin Pty Ltd
9 Atchison Street, St Leonards, NSW 2065 Australia

National Library of Australia
Cataloguing-in-Publication entry:

Framing technology: society, choice and change.

  Bibliography.
  ISBN 1 86373 525 9.
  1. Technology—Sociological aspects. 2. Technology—Social
  aspects. 3. Social change. I. Green, Lelia, 1956– . II. Guinery, Roger.

303.483

Set in 10/11pt Adobe Garamond by DOCUPRO, Sydney
Printed in Malaysia by SRM Production Services Sdn Bhd

10 9 8 7 6 5 4 3 2 1

For David Lake

CONTENTS

PART III FRAMING THE GLOBAL

FIGURES AND EXHIBITS

Figures

Exhibits

ACKNOWLEDGEMENTS

Intellectually this book reflects seven years' participation in the Australian Communication Association and continuous exposure to the two refereed journals it sponsors, *The Australian journal of communication*, and *Media information Australia*. To the journals' editors and contributors, especially to Roslyn Petelin, our appreciation and gratitude.

We owe a special debt to Brian Shoesmith and Robyn Quin who first encouraged Lelia's work in communication technologies; to Warwick Blood and Len Palmer at Charles Sturt University—and members of the MA (Communications) course; and to Bob Hodge, Tom O'Regan, and Toby Miller, all (at some time) of Murdoch University. David McKie, Michael O'Shaughnessy and other colleagues at Edith Cowan University—in Media Studies, Library and Information Sciences, Media Services and in the Faculty of Arts—have offered feedback and encouragement and enabled Lelia to take advantage of (greatly appreciated) professional development leave.

Four cohorts of Edith Cowan University's media, library and information science students helped shape this text which arises from their studies. We cannot thank Ann Manning and Debra Westerberg enough for holding the threads together when diverse frustrations threatened to unravel them. Della Commons juggled the discs, while David Rebbettes rescued us when we crashed and helped with the glossary. Eric McDonald met our impossible copying deadlines. Thank you, Judy Clayden, for the index and Elaine Pattullo for the copyright permissions. We also acknowledge the help of Lynda Davies, Jimmy Green, Wayne Harvey, David McKie, David More, David O'Brien, David Rebbettes and other contributors who have made comments and suggestions about the Abbreviations and Glossary.

Elizabeth Weiss at Allen & Unwin shepherded this book from its inception. She has been generous in her support and sparing with her criticism. Rebecca Kaiser guided the production process. Joan Smith, on another continent, sustains the desire to publish further.

Carmen and Benedict, Carmel, Pam and Marge—they all know what they did, and that no-one else could have done it for them. A special thank you also to Paddy and Jimmy Green for fielding the faxes and fuelling enthusiasm just when it was needed most. We promise not to put them, or Betty, through it again . . . for the moment.

David Lake was a wonderful librarian and a better friend. His example is an inspiration to those who knew him and this book is dedicated to his memory.

Lelia Green
Roger Guinery
Perth, Western Australia

CONTRIBUTORS

TREVOR BARR is an associate professor and head of The Media and Telecommunications Centre at Swinburne University of Technology. Trevor has worked extensively with the Commission for the Future, is a regular media commentator, and is a consultant for business—and the government—identifying industry opportunities in the telecommunications sector and recommending appropriate policies to meet the challenges ahead. Trevor is author or editor of three books on the media in Australia.

DICK BRYAN is a senior lecturer in economics at Sydney University, specialising in international trade and finance and Marxian economic theory. He is director of the University's Economic and Regional Restructuring Research Unit. He is currently completing a book analysing the contradictions for nation-state policy inherent in the international mobility of capital.

ROGER CLARKE is reader in information systems in the Department of Commerce, Australian National University. Roger's primary interests are in application software technology and its management, and in organisational, economic and legal aspects of information technology. Roger has been involved in the information privacy debate since the early 1970s and is Director of the Australian Computer Society's Community Affairs Board and Vice-Chairman of the Australian Privacy Foundation.

LYNDA DAVIES is an associate professor in information systems at Griffith University, where she has worked since her arrival from England in 1989. She researches in the area of electronic communities and organisational communication. Together with Wayne Harvey, Lynda is developing a conceptual framework of networking to elaborate theory and provide guidance for improving networking practices.

MICHAEL GALVIN is a senior lecturer in communications at the University of South Australia. His current research interests include the application of narrative theory to established and emergent forms of media, including computer-mediated communication, and the implications of technological change for participation in private and public life.

LELIA GREEN lectures in media studies at Edith Cowan University.

Concentrating upon research into the social dimensions of communication technologies, she has consulted for a variety of government and business organisations. Lelia is an active member of the Australian Communication Association and first became interested in society and technological change during research into communications in remote Western Australia.

ROGER GUINERY is a specialist in policy development and the management of change. He has consulted to the Western Australian Office of Communications, and to Golden West Network, on the needs of remote and regional Australians. Roger has played a leading role in the introduction of electronic trading into the Western Australian public sector.

WAYNE HARVEY is a researcher and post-graduate student in Information Systems at Griffith University. He brings to his work extensive technical knowledge and expertise acquired outside the academic context. Together with Lynda Davies, Wayne is working on the development of conceptual models as part of his research into electronic communities and organisational communication.

ADRIANNE KINNEAR is a senior lecturer in the Department of Science at Edith Cowan University. Her research interests are in the areas of the ecology and identification of soil fauna and sustainable technologies for agriculture. Adrianne's teaching extends into the areas of biotechnology, environmental sustainability and the identification and analysis of appropriate technologies.

DAVID McKIE, a Scot by birth and a Western Australian by choice, teaches in the Media Studies Department at Edith Cowan University. Obsessed with paradigmology in cultural studies, popular science and life, his research interests include popular genres such as science fiction, crime thrillers by women and sitcoms. He also writes on forging national identities, postcolonialism, tarot, ecology, pre-modern (especially non-Western) philosophies, postmodernism and how the last four might connect.

ALBERT MORAN is a senior lecturer in the School of Film and Media at Griffith University and has published nine books on Australian film and television. Albert is currently researching into men in the cinema, and into aspects of export strategies used by Australian television production companies.

DAVID MORE holds degrees in science and medicine, a medical research doctorate and fellowships in anaesthesia and intensive care medicine. Director of Health Services Consulting for Ernst & Young Australia, David previously worked as deputy executive director, information systems, for the NSW Department of Health. His early career was in intensive care medicine and medical research culminating in the position of director of accident and emergency services at Royal North Shore Hospital, Sydney.

ELIZABETH MORE is a professor of management, deputy director of, and director of research at, Macquarie University's Graduate School of Management. Elizabeth has particular research interests in organisational communication and culture, media management, and communications technology and policy.

SUSAN OLIVER is a past managing director of the Australian Commission for the Future, and works to promote a longer term perspective in the decisions and policies of business and government. A publisher, author and strategic planner with a particular interest in technology, Susan has managed research and consulting businesses and sits on the advisory boards of information and communication studies of three Australian universities.

LEN PALMER teaches sociology, communication and cultural studies at Charles Sturt University in Bathurst. After a career in technical telecommunications, he took a PhD in sociology from La Trobe University. With a long-term interest in technology, Len is currently researching the cultural contexts of electronic mail, and is interested in popular culture—especially music, video and computer games; and social theory—especially feminism, hermeneutics and theories of modernity and postmodernity.

JULIANNE SCHULTZ is a visiting fellow and research affiliate in the School of Social Sciences at the Australian National University. Previously she was an associate professor at the University of Technology, Sydney. Julianne is the author of *Steel city blues* (1985), *Accuracy and Australian newspapers* (1990), co-author of *The phone book* (1983), and editor of *Not just another business* (1994). She has worked as a journalist, journalism educator and media researcher.

JUDY WAJCMAN is an associate professor in Sociology at the University of New South Wales, Sydney. She is currently principal research fellow at the Industrial Relations Research Unit, University of Warwick, UK. She is the author of *Women in control* (1983), and *Feminism confronts technology* (1991), and co-editor of *The social shaping of technology* (1985).

ABBREVIATIONS AND GLOSSARY

This glossary is a starting point to help readers understand passages in which unfamiliar words appear. It makes no claim to be a collection of comprehensive definitions. Please refer to specialist subject directories and dictionaries for a detailed description of technical terms. It may also be useful to follow concepts through the index of this, and other, books and to cross-reference different authors' usages. If a term is explained in the text, in the only chapter in which it occurs—or in the *Concise Oxford Dictionary*—it has been excluded.

AHM: Australian Health Ministers

AARNET: Australian Academic Research Network

ABA: Australian Broadcasting Authority

ABC: Australian: Broadcasting Corporation

ABT: Australian: Broadcasting Tribunal

AGROSYSTEM: see Chapter 14

ALGORITHM: step-by-step logical sequence which embodies the rules for interpreting the information it contains. Different algorithms serve different purposes. When a communication is *digitised*, it is encoded algorithmically. Digital data travels in algorithmic form, and algorithmic rules dictate decoding back to the original form at the destination, for example to computer data, sound waves.

ANALOGUE: a representation of information which is capable of an infinite number of levels. To be distinguished from *digital* representations of information which, depending on the way it is coded, can only represent a fixed number of levels. An example is the analogue watch, on which the hands move continuously with the passage of time compared with the digital watch which clicks over the seconds, minutes and hours.

AOTC: Australian and Overseas Telecommunications Corporation

ARTIFICIAL INTELLIGENCE: a capability possessed by some inanimate objects to replicate some aspects of the intelligent behaviour usually seen as being reserved to human beings and some other advanced animals. The capacity for artificial intelligence is typically provided by a computer program which operates, as an example, a robot.

ASSEMBLY CODE: a language used to program a computer at a very basic level. To be distinguished from more advanced computer languages by the fact that an assembly language is unique to each type of computer, and usually contains only the simplest type of instructions within it; for example, add this to that and place the result here, store this data element there.

AT&T: American Telephone and Telegraph

BBC: British Broadcasting Corporation

BIO-: something which is biological (organic), and alive—a result of natural processes. *Bioengineering* is the reconfiguration of natural building blocks into an artificial, but living, structure; with *genetic engineering* a specialism which operates at the level of the *chromosome*. *Biotechnology* is a general term to encompass all technologies involved in the manipulation of biological, that is living, material.

BROADBAND ISDN: see *communications technology*

BROADCASTING SPECTRUM: the spectrum of *frequencies* upon which radio and television signals are broadcast. When radio, television and voice were transmitted using *analogue* (*wavelength*-related) communications, the broadcasting spectrum was viewed as a finite resource which became clogged by too many users competing for too few frequencies, compromising signal quality and leading to interference (for example, of broadcast radio by police/taxi transmissions). *Digitisation*—with *multiplexed* signals, *algorithmic* encoding and using fibre optics to deliver signals—potentially permits an infinity of channels.

CAD/CAM: Computer Aided Design/Computer Assisted Manufacture

CAPITALIST EXPANSION: the tendency of capitalism is to expand, because competitive pursuit of profit (and capital accumulation) involves the creation of new markets, and of new products for old markets.

CAT SCANNER: a diagnostic imaging tool that uses X-Rays and a computer program to create cross sectional images of body parts. CAT stands for computerised axial tomography.

CBNs: computer-based networks

CD-ROMS: compact discs–read only memory

CHROMOSOME: the elements within a living cell which contain the genetic information specific to that cell. The chromosomes are usually found in the cell nucleus and are the elements that are manipulated in the science of *genetic engineering*.

CRC: chlorofluorocarbon

CNN: Cable News Network

COMMUNICATIONS TECHNOLOGY: technology used for communicating—but also storing, retrieving and packaging (manipulating)—information. Such technologies include the pencil, telephone, computer and satellite. The *communications revolution*, the progressive separation of communication from transport (see Chapters 3, 12), is an integral part of the *information revolution.* An *intelligent communications* system has *artificial intelligence* built in, monitoring the flow and quality of the information communicated and making sure that it reaches the destination in the correct form (see Chapter 7). *Broadband ISDN* (Integrated Switched *Digital* Network) is an example of such a system. The 'broadband' prefix indicates that the system is designed to include many channels (bands), including multi-media transmissions. ISDN pathways utilise a variety of communications media—for example, satellite and/or *fibre optics.*

CONSTRUCTION OF MEANING: building upon theories about language, this asserts that people arrive at understandings that are created by language. Meanings are not 'out there' waiting to be picked up, they are a product of the way people express themselves, and are constructed by language. This viewpoint suggests that there is no certainty, no neutrality and no *objectivity.*

CPSR: Computer Professionals for Social Responsibility

CROSS-MATCHING: see Chapter 9

CROSS-SUBSIDY: the use of profits from one sector to subsidise an unprofitable sector, usually to permit the possibility of universal service at an affordable price. (See Chapter 6)

CSOs: community service obligations

CULTURAL STUDIES: an area of interdisciplinary study which assigns value to popular culture and promotes a more democratic society with particular respect to class, race and *gender.*

CUSTOMER SERVICE: see *public interest*

CYBER: something constructed artificially using *digital* technologies; thus *cyberspace*, a digital construction of space; and *cyborg*, a meld of digital/organic (often conceived of as actually living machines)

DATAVEILLANCE: see Chapter 9

DECONSTRUCTION: a critical approach to a text which subverts the power of the author, and undermines attempts to pin texts down to a single meaning by considering the gaps; what the text does not say. (Also see *structuralism*)

DEREGULATION: see Chapter 6

DETERMINISM: the notion that something is so powerful that it is beyond human control, for example, in *technological determinism.*

DGT: Direction Générale des Télécommunications—the French Government department with oversight of telecommunications policy and practice.

DIAGNOSTIC IMAGING: the use of technology to create images of body areas beyond human sight as an aid to diagnosis and treatment. Ultrasound uses sonic waves beyond human hearing; magnetic resonance imaging (*MRI*) uses magnets to help identify details about masses within the body; while computerised axial tomography *(CAT) scanning* uses X-rays to create cross-sectional images of organs and soft tissues in the body—more than simple bone mass.

DEC: Digital Equipment Corporation

DIGITISATION: the rendering of information in *digital* form. In *analogue* phone services the sound is modulated at the mouthpiece, retaining the pattern of sound waves until it is converted back into waves at the earpiece of the listener. Digitised sound is encoded algorithmically, using on/off states, and allows compacting of the signal with consequent benefits in terms of carriage capacity, and cost.

DISCOURSE: a discourse provides a set of ways of expressing meanings and values in a given arena. It is the product of power relations in society and sets the parameters for the *construction of meaning*. (See also *feminism* for an example.)

DNA: The molecule deoxyribonuceic acid which has the chemical bases for the genetic code. See also *genetic engineering* and *chromosome*

DUOPOLY: see *oligopoly*

E-MAIL: *electronic* mail, the use of computers in *LAN*s or *WAN*s to communicate privately with individuals, or publicly, using bulletin boards.

ECI: entertainment, communication and information

ECOSYSTEM: see Chapter 14

EFFECTS RESEARCH: a branch of audience/readership research which attempts to describe and quantify media effects such as the effect of violent television images upon children, or of pornography upon sex-offenders. There is a tendency to assume that there is an effect (see *construction of meaning*), although most effects described are subtle and highly conditional.

ELECTRONIC AND MICROELECTRONIC: applied to *microchip* technologies where the electrical circuits have been progressivly miniaturised to allow quantum leaps in available processing capacity while minimising physical size; often used to indicate some element of in-built computer control as in *robotics*.

ETHERNET: one of a number of *protocols* that are used to transport digital information between computers. Such basic information flow is a fundamental technology for the creation of computer networks.

ETHNOGRAPHY: the study of patterns of living of a series of individuals, or of a group of people. Tends to provide large amounts of data, which are affected by the presence of the observer/researcher, and which reflect their priorities.

FEMINISM: a *discourse* which assumes that *gender* is a political as well as social construction, and that for women to empower themselves involves them in a struggle against dominant masculine and patriarchal *paradigms*.

FEV: front-end verification

FIBRE OPTICS: the technology of using fine glass fibres to carry laser light which has been modulated in such a way that information can be transferred from one end of the fibre to the other. Laser energy travels at the speed of light and its power is concentrated as a result of the photons all having the same *wavelength*. *Digitisation* and *multiplexing* allow optic fibres to carry immense amounts of information.

FOURTH ESTATE: journalists and/or the media/press. The term acknowledges their independent political importance. The three estates which made up the estates-general in France before the French Revolution were: nobility (first estate), clergy (second estate), commoners (third estate). The press claimed authority as an independent but essential element of the political process in the nineteenth century, hence also the 'electronic estate' in the twentieth century.

FREQUENCY: the point on the *broadcasting spectrum* at which that signal is transmitted and can be received. The frequency of a radio signal, for example, is related to its *wavelength*. Longer wavelengths travel at lower frequencies. The frequency (the number of waves per second) is measured in hertz: Hz. A range of neighbouring frequencies make a spectrum, such as that of visible light (each colour has its own frequency).

GATT: General Agreement on Tariffs and Trade, a multilateral trading agreement which operates in conjunction with bilateral and regional trading agreements.

GENDER: a social *construction* with political implications, justified by biological difference. Sex may be decided biologically, gender is determined culturally. Conceptualisations of biological difference are also social constructions. (See *feminism*)

GENETIC ENGINEERING: artificial manipulation of the *DNA* which makes up *chromosomes*. It occurs naturally when an organism mutates.

GENRE: an identifiable form (or *discourse*) which allows *texts* to be categorised according to whether they conform to an appropriate *paradigm*, for example, detective fiction includes clues, and romances have a happy ending. A text need not be written, or be a media product; it can be any human/social *construct*—a life, Peter Garrett's say, or a city, for example, Perth. To view such artefacts as texts, however, a *reader* would normally subscribe to a *cultural studies*-type discourse, and use appropriate discursive practices.

GLOBAL/ISATION: the notion that a networked world is interconnected rather than divided—by satellites, by *fibre optics*, by *digital* information—and that the whole is greater than the sum of the parts. The concept is usually associated with such terms as *vectors*, descriptors of information flows and information power within the global context; and the term *global village*. The argument is that better communications reduce the relevance of physical distance and, controversially, of national difference, through networking global participants as fellow villagers. Such analyses tend to forget that only the rich are networked and that the *information poor* are effectively invisible.

HDTV: a high definition television that is more than a better TV set. It is a higher resolution, panoramic-image system. Competing methods of achieving HDTV are battling it out nationally/internationally as the videorecording formats—VHS, BETA and Philips—did. The marketing battle is further complicated by the different network broadcast systems *PAL, SECAM* and *NTSC.*

HEGEMONY: the process by which the subordinate are encouraged to consent to a system that subordinates them. This would be achieved when they agree to view the system (capitalism, patriarchy, *technological determinism*, etc.) and its everyday workings as common sense. It suggests the dominance of one power/state/*discourse*/idea within a given arena.

HERMENEUTICS: the creation of interpretations; the act of interpreting.

HMDs: Head Mounted Displays

HYPER: concentrated, essential, distilled, exceeding. A signal that the writer wants to add a gee-whiz element, for example, *hypereternity*—eternity plus—and *hyperreality*—reality plus! Tends to mean that this discourse involves an artificial, imaginary or constructed hyper-something.

ICAC: Independent Commission Against Corruption

IMPERIALISM: associated with the domination by one culture or country— the colonial power—of other countries or cultures—the colonies. Involves the enrichment of the imperial power by the impoverishment of the oppressed. *Post-colonial* analysis starts out with the idea that any *text* produced by a previously colonised society, or by a member of that society, fundamentally reflects the experience of oppression, that is, meaning remains the *construct* of the *reader* not the text-creator.

INFORMATION OVERLOAD: the inability of the information rich to handle the volume of information available—leads to specialised economic sectors of *information packaging*, tailoring and manipulation.

INFORMATION REVOLUTION: the process by which the material base of industrialised nations became grounded in an *information economy*. Conceived broadly this can include libraries, education, law, administration, media etc. *Information societies* were originally defined as over 50 per cent dependent upon information, and were sometimes referred to as *post-industrial* societies, because the industrial base was no longer judged to be the prime constituent. The use of *information technologies*, especially computers, databases and telecommunications (see *communications technology*) means that information societies are characterised by a knowledge explosion with the amount of information in the (developed) world estimated to be doubling every five years.

INFORMATION RICH/POOR: as the knowledge explosion accelerates, those with access to the information produced—the information rich—become richer. At the same time the *information gap* widens between the information rich and poor. (See *globalisation*)

INSTRUMENTALIST: associated with the 'technology is *neutral*' position. Suggests that technologies have no *meaning* or relevance other than as tools.

INTELLIGENT COMMUNICATIONS: see *communications technology*

INTERACTIVE: the capacity of a communication medium to be altered by, or to have its products altered by, a user or audience. A technology that requires input from a user to work effectively, for example automated teller machine, telephone voice messaging.

INTERESTEDNESS: the conceptual baggage which necessarily contaminates the activity of analysis. This precludes people ever being *objective*, or able to achieve analytical neutrality. It is a defence against bias for a person to be honest with themself and their audience about the *discourse/s* being used, and about the interests implicit in their position.

INTERFACE: a combination of computer hardware and software which permits the passage of information from one computer system to another. Interfaces may be either unidirectional (one way) or bi-directional (both ways). Sometimes used to refer to the hardware and software processes by which computer users utilise the technology.

INTERNET: a huge global network of computer systems which provides users with access to electronic mail, a range of information sources (data bases) and mechanisms by which collections of information (computer files) can be moved between users. *AARNET* is part of the internet.

INTERPRETIVIST: in opposition to *positivist*. Positivism argues that there are observable facts and phenomena, and that these are the foundation of knowledge. *Interpretivist* viewpoints counter by arguing that there is no *objective* observation—of facts or phenomena—and that everything should be acknowledged as *subjective* interpretation. *Discourse* analysis is an implicitly interpretivist tool. To be *post-positivist* is to reject explicitly the epistemology (theory of knowledge) of positivism.

IPD: intrapenile device

ISDN: see *communications technology*

IT: information technology

ITU: International Telecommunications Union

JSCAC: Joint Select Committee on an Australia Card

KNOWLEDGE PARADIGM: see *paradigm*, and Chapter 10

LOCAL AREA NETWORK (LAN): a network of computer systems that are confined to a single site or campus (compare with WAN—wide area network). LANs are often connected to other LANs using a wide area network. In this situation there is usually a gateway from each of the LANs to the WAN.

LEAN: Law Enforcement Access Network

MARKET ECONOMY: driven by capitalism, by the quest for profit. (See *capitalist expansion*)

MEANINGS: see *construction of meaning*

MEDIA DETERMINISM: see *technological determinism*

MEDIA DIVERSITY: used both to mean a variety of media, for example, print, television, radio, and a range of different products and viewpoints within a given medium, for example, a choice of newspapers expressing different viewpoints, including minority opinions and ideas which subvert the interests of elites.

MERKIN: female pubic wig used in burlesque shows as a defence against the accusation that women were unclothed; also used in early theatre by boys/men acting female parts.

MHR: Member of the House of Representatives

MICROCHIP: see *electronic* and *microelectronic*

MODEM: see *internet*

MONOPOLY: see Chapter 6

MORPHING TECHNIQUES: manipulation of image-elements to artificially change one image into another.

MRI SCANNER: a diagnostic imaging tool that uses the behaviour of the body components in a strong magnetic field, combined with a computer, to create images of the internal parts of the body. MRI stands for magnetic resonance imaging.

MULTI-MEDIA: the use of *interactive* multiple media components that respond to inputs, for example, a computer program through which a user can trigger text, graphics, pictures and sound.

MULTIPLEXING: the interweaving of communication elements

MYTH/OPOETICISM: the creating of myths, shared social understandings that help inform members of a cultural group as to the nature of that group, and to unite the group, for example, in the creation of a *national identity*. Sometimes associated with the suspension of disbelief in something which would usefully be true, for example, equality between the sexes might be seen as mythic. Myths often work below the level of consciousness. Stories, *fictions*, narratives and legends frequently serve mythic ends.

NAPLPS: the name of the algorithm for the computer-generated picture and word interactive system used by the Native American networked nations.

NARRATIVE THEORY: *readers* bring to an event or experience a sense of before and after required to understand the story. If an event is *unnarratable* it is beyond the context of the rules and understandings which society uses to *construct meaning*.

NATIONAL IDENTITY/IES: more than flags, anthems and coinage, national identity rests upon *myths*, for example, the bush, pioneer women, yuppies. The meaning of belonging to a national group (see Chapter 3) is often revealed through analysing (or *deconstructing*) the myths and narratives which help *construct meaning* for that group.

NSWPC: New South Wales Privacy Committee

NTSC: colour television broadcasting system developed in the US by the National Television System Committee.

NUCLEIC ACID: see *DNA* and *genetic engineering*

NWICO: New World Information and Communication Order

OBJECTIVITY: see *construction of meaning*

OECD: Organization for Economic Co-operation and Development

OLIGOPOLY: market domination or sharing by more than one company. (Two players are sometimes termed a *duopoly*.)

OTC: Overseas Telecommunications Corporation

PAL: Phase Alternation Line system, developed in Europe and used in Australia, New Zealand and the UK.

PARADIGM/ATOLOGY: the study of frameworks of understandings within which *discourses* are constructed and communicated. Analysing a paradigm can be the first step to resisting or subverting the *hegemonic* power of the elite whose interests that paradigm serves.

PIN CONNECTIONS: the standardised connections between the wiring in a cable and the terminal plug.

POSTCOLONIAL/ISM: a double focus where the first focus concentrates on the processes by which cultures are affected by *imperialism* from the moment of colonisation to the present. The second focus conceives of post-colonialism as a set of discursive practices clustered around resistances to colonisation, colonialist ideas and their contemporary legacies.

POST-INDUSTRIAL: a term for the new social forms emerging in post-1945 advanced industrial societies. Post-industrial societies are characterised by an information revolution and commentators debate the existence of features such as the leisure society, the shift from industrial production to a service economy, and the death of the working class.

POST-STRUCTURAL/IST: see *structuralism*

POST/POSITIVIST: see *interpretivist*

POSTMODERN/ISM: tends to be characterised as a period distinct from modernity, although there is much argument about the nature of the distinctions. Broadly seen as reflecting recent socio-economic ('late capital') developments, and/or an information (not industrially) ordered society, the postmodern comprises a cluster of features, but especially loss of faith in modernity's certainty regarding history, linear progress and universality. Postmodernism celebrates differences and encourages representational practices which express differences. (See also Chapter 12)

PRE-MODERN: peasant, pre-industrial (See Chapters 12, 14)

PRIVATISATION: the passing into private ownership, usually by share float, of publicly (government) owned industries.

PROSTHETIC: technological device used in or by a living creature to enhance or replace a natural function or skill.

PROTOCOL: a series of 'internal' commands by which a sender and recipient machine can inform each other of the status and destination of a specific data set which is to be transferred between the two machines.

PUBLIC INTEREST: when an aspect of a debate requires consideration of wider social costs and benefits than company profitability or government efficiency. *Public service* privileges the needs of the public regardless of their financial status; *customer service* considers the needs of people rich enough to be customers. The introduction of *user-pays* changes a public service into customer service.

R&D: Research and Development

RCCT: Randomised Controlled Clinical Trial

READER: a term making visible the active power of the person constructing *meaning* and interpreting a *text*.

RENAMING: see Chapter 10

ROBOT/ICS: either machines that perform physical movements and have a computational ability, or computers that have the characteristics and physical movements of an action-oriented machine. A robot has some *artificial intelligence*—the capacity to sense, select and input relevant data and possibly to react to events and perform actions in response.

SAMIZDAT: unofficial, illegal, resistant media generally opposing governments of totalitarian regimes.

SECAM: French-developed system of colour television, *séquence à mémoire*.

SECOND WORLD: the previous communist bloc in Eastern Europe. First World refers to rich, industrialised, developed and powerful nations while the Third World refers to the poor, lesser-developed or developing countries often powerless.

SHIELDING: conductive shield placed around cabling to reduce data transmission errors caused by external electrical radiation (noise).

SMOS: Special Minister of State

SOCIAL DETERMINISM: sees technology as expressing the priorities of the social elites which create or utilise the technology. (See *determinism*)

SOVEREIGNTY: a (*mythic*) idea that a nation is independent, and in control of its own future. (See Chapter 11)

STORIES: social constructs that explore or explain some element of life, often serving a *mythic* purpose.

STRUCTURALISM: originally challenged intellectual orthodoxy by denying commonsense and/or essential meanings (apart from cultural context). Semiotics, a branch of structuralism, is of continuing importance to *cultural studies*. Structuralism seeks to identify common elements of social practices or cultural products and the rules which combine them. *Deconstruction* goes further to identify missing elements in a *discourse*, for example, the *information poor* in the discourse of *globalisation*. *Post-structuralism* builds upon structuralism—often using psychoanalytical tools—to consider external structures, such as marriage, *gender* etc.

SUBJECTIVE/ITY: Cultural Studies tends to view people as 'subjects' constructed by language and institutions rather than as autonomous individuals. People derive identity through available linguistic and social categories such as *gender*, nationality, religion and education rather than creating themselves from their genetic inheritance and individual psychological makeup. (See also *interpretivist*)

SUSTAINABLE: see Chapter 14

TCP/IP: a common networking *protocol* originating in the UNIX operating system environment but now available in other operating systems as well.

TECHNOLOGICAL DETERMINISM: technology as outside social control, determining future social development and direction (see Chapter 1). *Media determinism* is a specialist application of technological determinism to communication media (see Chapter 13).

TECHNOLOGICAL/TECHNICAL DYNAMISM: acceleration of technological change, and the merging of new technologies to form novel products and resources.

TECHNOSYSTEMS: see Chapter 14

TELECOMMUNICATIONS: communications at a distance, usually implying the use of cable, satellite, *broadcasting spectrum, fibre optics* etc.

TEXT: see *genre*

TFN: Tax File Number

TNCs: transnational corporations

TOKEN RING TOPOLOGY: a networking *protocol* which performs similar functions to *ethernet* but uses a different technological approach. The ring of connected machines passes information in a circular fashion. The recipient of the data packet decodes the 'token' in the message to determine if it is the intended final recipient. If it is the destination, it accepts the message and removes it from the network, otherwise it passes the entire message on to the next machine in the ring.

TPE: Therapeutic Plasma Exchange

UNESCO: United Nations Educational, Scientific and Cultural Organization

UNIX: a multi-user, multi-tasking operating system used by many powerful computers. Unix contains many sub-programs that perform network operations transparently between several machines.

UNNARRATABLE: see *narrative theory*

USER-PAYS: see *public interest*

VECTOR: see *globalisation*

VIRTUAL REALITY, VR: see Chapter 2

WAVELENGTH: electromagnetic waves travel out in every direction from the point where they are generated. The distance between the peaks of the waves (where the electrical energy is strongest) determines the wavelength. Wavelength is related to *frequency*: the longer the wavelength the lower the frequency; the higher the frequency the shorter the wavelength. (See also *broadcasting spectrum*)

WCED: World Commission for Environment and Development

WIDE AREA NETWORK (WAN): a network where the machines are physically distant and are connected by a leased data communications line, microwave link or similar, for example, *internet*.

References

O'Sullivan, T. et al. 1987, *Key concepts in communication*, Routledge, London

Telecom 1986, *The Telecom directory of abbreviations, acronyms, jargon and technical terms encountered in Australian telecommunications*, 2nd edn, Telecom Australia, Melbourne

INTRODUCTION
Lelia Green

Any consideration of technology requires a working definition of what technology is before developing an understanding of its relevance. Technology includes ways of doing things as well as the machinery with which things are done. The layers of meaning offered by Judy Wajcman—who defines 'technology' as including objects, activities and knowledge—indicate that the term can refer to a transnational corporation (TNC), to a baby bottle, to canned beans and to human language. Three decades ago Marshall McLuhan noted 'It can be argued, then, that the phonetic alphabet, alone, is the technology that has been the means of creating "civilized man"—the separate individuals equal before a written code of law' (McLuhan 1964, p. 84).

This inclusive definition of technology is deliberately used to widen current debates about technological change and technology choice. Such a perspective rejects the argument that an individual needs a detailed understanding of assembly code to have an opinion on the cross-matching of databases, or of DNA to have a view on genetic engineering. It is no accident that the narrower the definition of technology, the more specialised the qualifications for taking part in the debate. Such restrictions place the power for developing future directions for society into the hands of a tiny number of people. In contrast a broader view asserts that technology choice should involve everybody.

Technology is the bedrock of our information society but public debates of technological change tend to concentrate on the so-called hi-tech: communication networks, virtual reality, the microchip. Where a myopic attention to physical objects is broadened to include social systems there is a tendency to concentrate upon technology, work and unemployment. Although useful, this is a restricted perspective which is well handled elsewhere (such as Jones 1990; Willis 1988). This collection reframes the discussion to include consideration of such technology choice as the use of language in media coverage of the Gulf War, and the TNC as an example of the technology of western capitalism.

Framing technology is an interdisciplinary activity and the breadth of such an enterprise is reflected in the scope of the essays included in this volume. Social scientists increasingly reject as inadequate notions of objectivity and analytical neutrality. What is seen is highly dependent upon who is doing the looking and the position from which they start. Perspectives on offer in this book complement and challenge each other and include sociology, environmental science, cultural studies, economics and management. Different disciplines offer different theories of technology—and imply different definitions—but taken together these chapters demonstrate how important it is to begin investigating technology from a variety of positions.

Pinning down the concept of framing, and of technology, is like trying to nail jelly. As the action begins to bite, everything moves. Unlike nailing jelly, however, there is a point to investigating relationships between frameworks and technologies. Understanding the framing of technological discussions reveals the interests served by the placing of the parameters.

At one level a frame operates as a showcase. These essays showcase contemporary views from a number of perspectives on key issues regarding technology. Secondly, framing also operates as a boundary within which a subject can be studied or appreciated, but by which it is safely constrained. Technology and technological debates are habitually framed by a small elite in western society using exclusive language and a narrow definition. This collection starts with the key question of who benefits (and who is disempowered) by the agenda being set in this manner: why are only selected people invited to the meeting?

Thirdly, technology has been framed by those who claim to judge it objectively, and who police the arguments concerning it. Is technology 'neutral'; an instrument of oppression (and surveillance, and capitalism, and war); or an expression of the society which creates it? It has been defended as innocent, attacked as guilty, and interrogated as an accessory after the fact. Simple verdicts miss the point. Who asks the questions, what are the acceptable topics? A key aim of this book is to encourage wider questioning. The answers lie in the asking.

Framing the individual

Judy Wajcman opens the debate by arguing that technology is not external to society, it is an expression of society. Technology cannot be reduced to hardware: it is meaningful only as a part of human activity, sustained by human knowledge. In particular, technologies are an expression of a masculine society, and it is male elites who determine what is (and what is not) framed as technological.

A feminist perspective on technology includes an appreciation of women's (hitherto hidden) contributions to technological development, and an understanding of the ways in which the masculine framing of technology frequently exploits women. Wajcman argues that a prime reason for understanding technological debates as a struggle for power between different groups is that only then is it possible to demand a part in the discussion.

David McKie develops the issue of the exclusion of women—and of indigenous peoples—when he considers the hip hyper-technology, virtual reality (VR). Addressing issues raised by VR, including cyborgs and genetic engineering, McKie argues that virtual worlds are almost exclusively created by white, middle class men (whether the venue is virtual sex or virtual cooking).

Concerned that the kinds of realities currently under construction will create unsustainable futures, McKie goes on to discuss evidence for complexity theory—itself being investigated through the medium of computer generated 'virtual life' forms. Throughout his chapter McKie suggests that the creation of a virtual reality is accompanied by moral choices, that environmental reality is not 'virtual' and that complexity theory—rather than cybersex—is the more likely framework within which to hope for the survival of the species.

Whether or not VR will 'eat television alive', TV remains crucially important to industrialised societies. Albert Moran's chapter focuses upon the broader cultural issues which concern television, as well as considering the political economy of broadcasting. A technological object is framed by the discourse within which it is discussed. For example, a car might be conceptualised in terms of fuel economy, or top speed, safety or price. These functional and economic agendas are familiar from advertising strategies, and tend to influence the way in which people in the west perceive domestic vehicles. Consider, instead, other cultural contexts which could encompass a car. For example, these are Igor Kopytoff's views:

> The biography of a car in Africa would reveal a wealth of cultural data: the way it was acquired, how and from whom the money was assembled to pay for it, the relationship of the seller to the buyer, the uses to which the car is regularly put, the identity of its most frequent passengers and of those who borrow it, the frequency of borrowing, the garages to which it is taken and the owner's relation to the mechanics, the movement of the car from hand to hand over the years, and in the end, when the car collapses, the final disposition of its remains. All of these details would reveal an entirely different biography from that of a middle-class American, or Navajo, or French peasant car (Kopytoff 1986, p. 67 cited in Silverstone & Hirsch 1992, p. 17).

Television is contextualised as part of an ECI (entertainment, communication and information) clan which claims the printing press as a distant ancestor and which enjoys sibling rivalry with computers, with compact discs and with video games. Moran foresees the end of television in its present form, with an increasing number of technological systems and services competing for the leisure (and ECI) attentions once exclusively devoted to the box.

Susan Oliver issues a challenge to the public: 'The community as a whole will have to become better educated about science and technology so, through understanding, they achieve greater control of it'. Exposing the narrow horizons of national goals which only address economic indicators, Oliver argues that policy-makers hark back to a past of limited relevance. Individual and community commitment to social equity, coupled with a vision for the future, are critical to the technological agenda. Oliver does not propose an education program relating to Ohm's Law or the Periodic Table; instead she advocates active discussion of possible social goals, such as those put forward by Ian Miles (1985). These include achieving a harmonious relationship between persons, society and nature and a living presence of the future.

Technologies of health care delivery are investigated by David More and Elizabeth More in 'Till death us do part: technology and health'. Why are the costs of health services escalating? Do more technological resources mean better health profiles? Who should decide what is acceptable spending upon technological resources, and what factors should be taken into account? Access to life-preserving dialysis machines, for example, is identified as an area in which there are already restrictions upon 'the very old'. As resources increasingly fail to keep pace with the growing demand for medical technologies, even the comparatively youthful may wish there was more general debate about ethical issues such as opting out of donating organs at death, rather than opting in.

Between them, these five chapters address issues which impinge upon the lives of all Australians and New Zealanders at some stage. A vision for the future, what kinds of virtual realities to work towards, medical procedures, television products, and the gendered framing of technological debates: these all have implications for individuals. Responses people make to these challenges will determine whether or not they take more power in making the technology choices which affect their lives.

Framing the communal

Moving from individual concerns to issues more firmly grounded in the public domain, the contributors in this section consider matters of relevance to communities—usually in terms of the nation-state, but sometimes on a more local or global stage. The actions and priorities of elected representatives in government (and their professional administrators), along with key members of 'the market place', circumscribe the bumpy communal pitches upon which technological issues are contested. Also central to the framing of technology debates, the media and communications industries serve as highly partisan umpires.

Len Palmer addresses the 'contestation, struggle and possible conflict' implicit in the regulating of technology. Unintended outcomes, and the impossibility of guaranteeing consequences, characterise technology choice. Yet models of technology regulation do not currently include provisions for intervention to modify or reverse decisions. Palmer borrows from developments in the social sciences which eschew notions of objectivity and rationality. In their place is an argument that all representations of reality—including objective ones—are constructed by people with a stake in the discussion. Palmer concludes that current models of technological regulation amount to unacknowledged social gambling.

In the international arena, concerted action by third world countries has shown up the political nature of regulating technology. These countries used the International Telecommunications Union (ITU)—a United Nations forum—to put forward demands for the fairer distribution of information and communication technologies, and for more equitable technology choice internationally. The United States and Britain withdrew from the forum rather than debate the issues. Palmer suggests that if Australians saw national decisions on technology choice as equally political there might be more debate on them—and more democratic input.

Trevor Barr argues that level playing fields are not necessarily the sign of a clever country. The free-market approach failed to foster an indigenous computing industry in the 1980s, and there are fears for the future of Australia's telecommunications sector in the 1990s. Further, analysing the trend towards globalised corporations, Barr warns against the prospect of there being no national broadcasting in Australia, although Star TV in Hong Kong might radiate the ABC and SBS by satellite throughout Asia. The commercial channels might no longer be networked from Sydney, but from the West Coast of the United States (although they may well be owned by Kerry Packer and Rupert Murdoch).

The biggest danger Barr fears is the redefinition of 'the public interest'. It was in the public interest, for example, that Telecom Australia used a cross-subsidy so that people in remote areas could afford a telephone service. Barr's fear is that the dynamics of competition and privatisation have included a conceptual sleight of hand. Public service has become customer service. The new orthodoxy is that what is good for the market is good for Australia.

Examining the relationships between technology and democracy, Julianne Schultz comments that democratic applications of technology are unlikely to be fully realised while commercial interests market technology for profit. Yet microelectronic technology has 'the capacity to affect the ability of citizens to function in a democratic society, and as a result the potential to influence the very nature of democracy itself.

Potentially, technologies could be developed and distributed to fulfil a democratic agenda—such as greater access to information—and could exhibit additional characteristics by adapting the technologies currently available to meet individual needs. Schultz offers Eric Michaels' (1986) study *The Aboriginal invention of television*, as an example of the way technology can be adapted democratically to empower and permit the communication of alternative viewpoints. Since corporate profit is the major beneficiary of technological innovation, and as the global media industry becomes less representative of minority and dissenting opinions, democratic principles may be overlooked.

As 1984 fades into the middle distance, 'Dataveillance' makes uncomfortable reading. One of the catch phrases to emerge from the Australia Card debate was 'in a democracy the people should scrutinise the government—not the other way around'. Instead, an Orwellian future is increasingly possible. Dataveillance is 'delivering *1984* . . . just a little late'.

Roger Clarke's strategy is to identify and discuss some of the key tools of dataveillance—such as the Tax File Number—and examine the context in which these applications of information technology are championed. He considers actual and potential control mechanisms which aim to ensure that 'organisations' practices do not sacrifice humanity in the search for resource efficiency'.

Although computer-based networks have been viewed as open and democratic—it is up to the individual how much they write and what they read—Lynda Davies and Wayne Harvey demonstrate that these appearances of openness are not the only representation of reality. Networks superficially involve the relocation of communication away from usual social cues of status-display and special prerogatives, but these distinctions are encoded instead within the networked communication process.

Davies and Harvey identify two audiences at whom current models of computer networking are aimed: technologists and managers. Both of these have vested interests in maintaining their power relationships. These interests are not served by opening the discourse to wider participation. 'In the hands of the knowledgeable, machines are the means for heightening the politics of knowledge, creating greater restrictions and more opportunities for political manipulation of others who are less knowledgeable.'

These five chapters on 'Framing the communal' address debates which are primarily circumscribed by the national arena. It is impossible to examine these issues, however, in isolation from worldwide trends. Whether it is the decline of the Australian computer manufacturing industry, self-determination in telecommunications, surveillance techniques, borderless data networks or transnational media conglomerates, national agendas are increasingly affected by international considerations. The collection of essays goes on to consider technology and global society.

Framing the global

In reframing the global picture, the conceptual requirement is to do more than simply pull back and widen the mental angle. The view-finder reveals a significant difference in the transition from communal to global considerations which is not there when moving from individual to communal issues. The global information society is greater than the sum of individual nation-state communities. It also encompasses a sense of placelessness.

Joshua Meyrowitz (1985) identifies four sets of meanings which comprise a 'sense of place':

sense	*place*
perception	social position
logic	physical location

Perceptions of social position have changed significantly over the past generation, particularly in terms of available gender roles. Perceptions of physical location may be more nebulous, but Meyrowitz argues that the electronic media is able to 'break down the distinctions between here and there, live and mediated, and personal and public' (Meyrowitz 1985, p. 308).

In being less firmly placed in an actual physical location, people find themselves moved by events which occur thousands of kilometres away, involving individuals they will never meet. In focusing more upon electronic communications, they engage less with their physical neighbours, and distance themselves from the joys and tragedies of people who live close by.

Along with changes in perception, argues Meyrowitz, come changes in logic. The logic of social position—the ways in which people understand individual roles and cues for appropriate behaviours: as daughters, mothers, sisters, sexual partners, union members, employees, church workers and political activists—has been critically affected by electronic media. The logic of physical place is experiencing a similar sea-change. Electronic media have:

> changed the logic of the social order by restructuring the relationship between physical place and social place and by altering the ways in which we transmit and receive social information. The changing relationship between physical and social place has affected almost every social role. Our world may suddenly seem senseless to many people because, for the first time in modern history, it is relatively placeless (Meyrowitz 1985, p. 308).

This sense of placelessness, emerging out of the globalisation of electronic communications media, is a new element in the information society mix. Networked corporations are not constrained by national boundaries or national contexts; as global, they are comparatively free from locational restrictions. These institutions challenge the traditional regulatory power of the nation-state and demand an international response. They are accountable only to their shareholders. Transnational corporations (TNCs) are particularly well qualified to act as the *bêtes noires* of the information age. Behind every technology alleged to be subverting democratic ends there is a TNC which uses the means of new communication technologies towards the ends of corporate profit within the global market. Herbert Schiller, a critical scholar, argues that TNCs support

> the existing global hierarchy of power . . . utilize the communication and telecommunication systems, locally and globally, to direct their complex and geographically dispersed operations. They have pressed for and obtained privatization of communication facilities in one national locale after another, enabling them to have the greatest possible flexibility of decision making and allowing them a maximum of social unaccountability (Schiller 1991, p. 21).

Transnationals use the technologies; they often manufacture the technologies; they frequently have power as a result of supplying the technologies:

> The good samaritan is no longer the peace volunteer who, furnished with a stock of slides provided by the cultural service of the embassy, reveals the grandeurs of the interplanetary race to the peasants of the Andes. He has become a salesman of the latest electronic models produced by the multinationals (Mattelart 1979, p. 77).

An apocryphal Coca-Cola executive is credited with coining the apologia that 'Coca-Cola is not a multinational, it's a multilocal', but such distinctions cut little ice with TNC detractors. Dick Bryan, however, argues that the problem does not lie with the TNCs—but with international capitalism. TNCs, although efficient organisational expressions of capitalism, are no more culpable than other large profit-making bodies, transnational or not. Nonetheless, the international expansion of capital has created new issues. It may not have subverted the nation-state, but it has demanded that nation-state policies recognise that the nation is but part of a global system. National policies can no longer be formulated as if governments have complete control over what happens within the nation.

'Missing the post(modern): cores, peripheries and globalisation' starts with the undoubted disparity in wealth and information technology between the first and third worlds. It goes on to discuss changes in communication technologies, and their link with the western experience of industrial revolution. Finally, the chapter examines 'soft imperialism', the role of popular culture in westernising—Americanising?—the rest of the globe.

Core/periphery distinctions, which inform much critical theory on technological disparities between the first and third worlds, are characterised by 'modern' analysis. This reflects a conceptualisation of industrial societies as having an industry-dense centre supplying manufactured goods to a rural hinterland. 'Missing the post(modern)' argues that the mind-set of postmodernism is the more appropriate one for conceptualising the importance of communication technologies to information societies.

The concept of a vector used to be associated almost exclusively with physics and mathematics where it denoted various dimensions—height, distance, speed and/or direction—of a moving object or force. Latterly the term has been used by McKenzie Wark (e.g. Wark 1991, following Paul Virilio), to address the nature and relevance of global communication channels. Michael Galvin argues that the power of the vector, which directs broadcasting, information and/or military intelligence in global networks, reached unprecedented heights during the 'Nintendo War'.

To the viewer/gunner involved in the Gulf conflict, the monitor images—phosphorescent-bright faces of doomed Iraqi troops—resembled a video simulation. The merging of reality with the semblance of Nintendo is one reason given by Galvin for a lack of journalistic comment; events were outside human experience and so became unnarratable. With the emphasis upon technology—targeting and weaponry—the estimated Iraqi death toll, put at between 500 and 1000 soldiers dead for every allied fatality, received little narrative attention. Galvin suggests that although the war received saturation coverage, people were underinformed—they lacked the information they needed to understand the human dimension.

Using an ecological perspective to examine the sustainable networks of the natural environment, Adrianne Kinnear underlines the fact that technology choice is a global issue. The accelerating spiral of consumption combines increasingly toxic waste and non-renewable energy source depletion with environmental costs which affect everyone. Kinnear's starting point in identifying ecologically appropriate technologies is sustainability. What are the features of sustainable technological development? Is it a goal worth recognising and working towards? What will the consequences eventually be of perpetuating the use of unsustainable technologies?

Ecological determinism is a focus increasingly important to the global technological debate but not often included within it. Kinnear addresses a variety of possible perspectives but suggests that future options are limited: 'Technocrats among us argue that radical directional change towards environmental sustainability will cause major social and economic dislocation. The ecologists argue that we have no choice.'

Although the environment, war, the media and TNCs are all global issues, they require an individual response—and individual stories—to make sense of them. The narrative structure will help determine the ending. In the last decade of the twentieth century there is an increasing need for everyone, but especially members of the information society, to decide for themselves the discourse they wish to construct around technology. In framing technology, people are framing the future society in which they will live.

References

Jones, Barry 1990, *Sleepers, wake!*, Oxford University Press, Melbourne

Kopytoff, Igor 1986, 'The cultural biography of things: commoditization as a process', *The social life of things: commodities in a cultural perspective*, ed. Arjun Appadurai, Cambridge University Press, Cambridge

McLuhan, Marshall 1964, *Understanding media*, McGraw-Hill, New York

Mattelart, Armand 1979, *Multinational corporations and the control of culture: the ideological apparatus of imperialism*, Harvester Press, Brighton

Meyrowitz, Joshua 1985, *No sense of place: the impact of electronic media on social behaviour*, Oxford University Press, New York

Michaels, Eric 1986, *The Aboriginal invention of television, Central Australia, 1982–6*, AIAS, Canberra

Miles, Ian 1985, *Social indicators for human development*, Francis Pinter, London

Schiller, Herbert 1991, 'Not yet the post-imperialist era', *Critical studies in mass communication*, vol. 8, no. 1, March, pp. 13–28

Silverstone, Roger, Hirsch, Eric and Morley, David 1992, 'Information and communication technologies in the moral economy of the household', *Consuming technologies: media and information in domestic spaces*, eds Roger Silverstone & Eric Hirsch, Routledge, London

Wark, McKenzie 1991, 'News bites: war TV in the Gulf', *Meanjin*, vol. 50, no. 2, pp. 5–17

Willis, Evan 1988, *Technology and the labour process: Australasian case studies*, Allen & Unwin, Sydney

PART 1

FRAMING THE INDIVIDUAL

TECHNOLOGICAL A/GENDERS:
TECHNOLOGY, CULTURE AND CLASS
Judy Wajcman

We tend to take for granted the world of things in which we live—a world of things that people have made. We do not ask why our refrigerator makes an annoying humming noise, nor why our domestic appliances are shaped the way they are. We think about electricity only when the bill has to be paid, or when the supply fails. An electric light bulb is an object that excites no comment. When technology does get into the news—as has happened with nuclear weapons, with the microchip, with test-tube babies—we often feel powerless to affect the course of events. People today are encouraged to take the view that while technological change has political effects, or requires a political response, it is not in itself the product of political forces.

But is technology just the result of rational technical imperatives? What if, rather than technology being neutral, it is the result of a series of specific decisions made by particular groups of people in particular places at particular times for their own purposes? And if it can be shown that political choices are embedded in the very design and selection of technology, then we can begin to think of technology as something we might shape consciously.

The single most influential explanation of the relationship between technology and society is the theory of technological determinism. According to this account, technologies themselves are neutral and changes in technology are the most important cause of social change. In this view, technology impinges on society from the outside. Although the scientists and technicians who produce new technologies are members of society, their activities are seen to be independent of their social location. Completely dedicated to the pursuit of knowledge and its practical application, they are represented as above sectional interests and politics.

Regardless of how and by whom it has been effected, some futuristic commentators claim, for example, that the microelectronic revolution is causing a new form of society to emerge. We will be forced to change our ideas of work and leisure, it is said, as the microchip 'puts millions out of work'. Social thinkers as diverse as Andre Gorz, Alvin Toffler and Barry Jones are promoting versions of the idea that changes in technology are bringing about a new 'leisure society' or 'post-industrial society'. For example, in *Sleepers, wake!* Barry Jones (1990, p. 254) argues that:

Technology can be used to promote greater economic equity, more freedom of choice, and participatory democracy. Conversely, it can be used to intensify the worst aspects of a competitive society, to widen the gap between rich and poor, to make democratic goals irrelevant, and institute a technocracy.

This statement implies that, while technology itself is neutral, the problem lies in its use and abuse by society or sections of society. As long as enlightened people are in control of the technology, all will be well. Taking this approach tends to blind us to the extent to which technologies in themselves have political qualities.

The economic shaping of technology

If technology is the result of neither scientific discovery nor an inner logic of development, what does shape technology? New technology typically emerges not from flashes of disembodied inspiration but from existing technology, by a process of gradual change to, and new combinations of, that existing technology. Even the so-called revolutions in technology turn out to have been long in the making. Existing technology is thus an important precondition of new technology. But it is not the only force shaping new technology. Technological systems are oriented to a goal and that goal is normally about reducing costs and increasing revenues.

Electricity supply systems, for example, have been private or public enterprises, and those who have run them have inevitably been concerned above all with costs, profits and losses. When technologists focus inventive effort on the 'inefficient' components of the system, for many practical purposes 'inefficient' means uneconomical. So technological reasoning and economic reasoning are often inseparable.

Thomas Hughes's (1985, pp. 39–52) work demonstrates this in the case of Edison's invention of the lightbulb. Edison was quite consciously the designer of a system. He intended to generate electricity, transmit it to customers, and to sell them the apparatus they needed to make use of it. To do so successfully he had to keep his costs as low as possible—not merely because he and his financial backers wished for the largest possible profit, but because to survive at all electricity had to compete with the existing gas systems. Significantly, Edison believed he had to supply electric light at a cost at least as low as that at which gas light was supplied. These economic calculations entered directly into his work on the lightbulb. A basic system cost was the copper for the wires that conducted electricity. Less copper could be used if these wires had to carry less current. Simple but crucial science was available to him as a resource—Ohm's and Joule's laws—from which he inferred that what was needed to keep the current low and the light supplied high was a lightbulb filament with a high electrical resistance, and therefore with a relatively high voltage as compared to current. Having thus determined, economically as much as technologically, its necessary characteristics, finding the correct filament then became a matter of 'hunt-and-try'.

The development of the common refrigerator provides another example. One hardly gives a moment's thought to the fact that we have electric rather than gas refrigerators. It is simply assumed that the electric model must be technically superior. Ruth Schwartz Cowan, an American social historian, has researched the rivalry between the gas refrigerator (the machine that failed) and the electric refrigerator (the one that succeeded). There were initially designs for both and, indeed, until 1925 gas refrigerators were more widespread. Cowan (1985, pp. 202–18) argues that electric refrigerators came to dominate the market as a result of deliberate corporate decisions about which machine would yield greater profit. The potential market for refrigerators, as well as the potential revenue for gas and electric companies, was enormous. By 1923 it was clear that every household in the United States could become a customer for a refrigerator. By the mid-1930s, most of the fundamental innovations in domestic refrigeration design had been made.

These innovations did not occur out of the blue—a great deal of time and money were spent in achieving them. Some manufacturers were going to succeed and others would fail. Large corporations, like General Electric, with vast technical and financial resources, were in a position to choose which type of machine to develop. Not surprisingly, with interests in the entire electricity industry, General Electric decided to perfect the design of the electric refrigerator. The refrigerator that General Electric introduced to the public in 1925 was the product of almost fifteen years of development work on the part of the firm's employees.

But what of the gas refrigerator? Why was it not developed? Cowan argues that the gas refrigerator had the potential to be a superb machine for household use. From the consumer's point of view, these refrigerators' chief advantages were that they were virtually silent; they were easy to maintain; and operating costs could be kept fairly low. Yet the machine, like the electric model, was going to require expensive development and promotion before it could be made commercially successful. The manufacturers of gas refrigerators, although they had a product with real advantages from the consumer's point of view, lacked the resources for developing and marketing their machine. So the demise of the gas refrigerator was not the result of deficiencies in the machine itself; rather, as Cowan argues, it failed for social and economic reasons. Her account illustrates that the household machines which we have, we have not because of their inherent superiority, not simply because of consumer preference, but primarily because of their profitability to large companies. Consequently, rather than having absolutely silent refrigerators in our kitchens, we have machines that hum.

Technology, economics and class relations

Paradoxically, then, the compelling nature of much technological change is best explained by seeing technology not as outside society, as technological determinism would have it, but as inextricably part of society.

What we think of as technology can be seen to have at least three layers of meaning, each of which involves social processes. At the most basic level, technology refers to a set of physical objects—such as lathes, vacuum cleaners or computers. To see these objects as nothing other than inanimate items is a very restrictive viewpoint and ignores the fact that they are developed, manufactured and marketed as part of economic and social activity. Further, a technological object such as a car is only a technology, rather than an arbitrary lump of matter, because it forms part of a set of human activities. A computer without programs and programmers is a useless collection of metal, plastic and silicon. So a second level of meaning acknowledges that technology refers to what people do, as well as the objects they use. Once we talk about people doing things, we are talking about society, and all the forces which direct and sustain human activities. Thirdly, technology refers to knowledge as well as to activities and objects. The knowledge is the social framework which informs and teaches the activities. It implies education, experience, basic competency and some perception of right and wrong ways of doing things. A driving test, for example, tests a complex set of knowledge and physical and social competencies, while the legal system—and the potentially tragic effects of making mistakes—encourage continuing observance of the rules. It is clear that the argument that technology is neutral, and somehow apart from society, ignores every important facet of the ways in which technology forms part of human life.

If, as we saw with the lightbulb and the electric refrigerator, technological systems are economic enterprises, and if they are involved directly or indirectly in market competition, then technical change is forced on them. If they are to survive at all, and even prosper, they cannot forever stand still. Technical change is made inevitable, and its nature and direction profoundly conditioned by this. And when national economies are linked by a competitive world market, technical change outside a particular country can exert massive pressure for technical change inside it. These simple but overwhelmingly important aspects of technical change were identified by Karl Marx in volume 1 of *Capital*, who attributed to them the unprecedented technical dynamism of the capitalist system. Marx's insights have been pursued less systematically than one might have expected. As Nathan Rosenberg has put it, 'Marx's analysis of technological change opened doors to the study of the technological realm through which hardly anyone has subsequently passed' (1982, p. viii). Economic calculation, however, is shaped by its social framework. Again, as Marx recognised, economic calculation and economic laws are not universal but are specific to particular forms of society. Even if in all societies people have to try to reckon the costs and benefits of particular decisions and technical choices, the form taken by that reckoning is importantly variable. Economic calculation presupposes a structure of costs that is used as the basis for that calculation. But a cost is not an isolated, arbitrary number of dollars. It can be affected by, and can itself affect, the entire way a society is organised.

This point emerges most sharply when we consider the cost of labour, a vital issue in technical change, because much innovation is sponsored and justified on the ground that it saves labour costs. In a capitalist society, a major factor affecting the cost of labour is capitalist–worker relations. In order to maximise profit, workplace technology may actually be designed to reduce the amount of control the worker has over the production process, by deskilling the workers and increasing the control of management. Labour historians have argued that, especially in the early phases of capitalist development, machinery was used by the owners and managers of capital as an important weapon in the battle for control over production. Thus direct relationships have been identified between particular arenas of industrial conflict and particular kinds of technical innovation.

Marx's classic account in *Capital* of the development of the automatic spinning mule in nineteenth-century Britain has, for example, been re-examined from this perspective by Bruland (1985, pp. 84–92). In the early production process of spinning the skilled adult male spinner had a central role. The spinners were highly unionised and their frequent strikes were a direct challenge to the power and profits of the cotton-masters. The self-acting mule was the employers' response to this threat. The explicit purpose of this invention was to break the power of spinners by replacing the men by the cheaper labour of women and children.

The machine tool industry in the USA provides a more contemporary example. In his book *Forces of production*, David Noble (1984) argues that a major goal of machine tool automation was to secure managerial control, by shifting control from the shop floor to the centralised office. Thus Noble argues that forces other than 'technical imperatives' were at work in the development of machine tools and that the form of automation was the result of deliberate selection for a particular purpose. Noble explains that there were at least two possible solutions to the problem of automating machine tools. Machining was in fact automated using the technique of numerical control. But there was also a prototype for a technique of automation called 'record-playback' which was technically as reliable as numerical control but enjoyed only a brief existence. Why, asks Noble, was numerical control developed and record-playback dropped?

It was the post-war period of labour militancy that provided the social context in which the technology of machine tool automation was developed. Record-playback was a system that would have extended the machinists' skill. Although the machines were more automated, the machinists still had the control of the feeds, speeds, number of cuts and output of metal; in other words, they controlled the machine and thereby retained shopfloor control over production. Numerical control offered a means of dispensing with these well-organised skilled machinists. The planning and conceptual functions were now carried out in an office because the machines operated according to computer programs. The machinist became a button pusher. Numerical control was therefore a management system, as well as a technology for cutting metals. It led to organisational changes in the factory which increased managerial control over production because the technology was chosen, in part, for just that purpose.

In fact, there is now a growing literature on the way in which the design of machines is shaped by class relations. However, people attempting to examine technological development as a social phenomenon are just starting to analyse the implications of gender relations, as well as other political and cultural influences. New technology is being introduced into a labour market in which women's labour is generally cheaper than men's. This may affect technological change in at least two ways. First, employers may seek forms of technological change that enable them to replace expensive unionised male workers with less costly female workers. Secondly, because a new machine has to pay for itself in labour costs saved, technological change may be slower in industries where there is an abundant supply of cheap women's labour. One explanation for the lack of technological change in the sewing process, for example, is the fact that women sew and have been available to work for low wages, either in third world countries or as immigrants to western capitalist countries.

Technology and gender relations

It is not only at an industrial level that technology is defined as a male preserve. Part of this construction of technology as masculine is the assumption that the domestic sphere is a technology-free zone. Yet the average household is packed with technology that women use, and men repair and design. White goods—such as cookers, refrigerators and washing-machines—like other forms of technology, are designed within a social context. The social prevalence of the private, single-family household run by an essentially unaided, usually female, housekeeper has profoundly structured the type of technology that is produced. As Ruth Schwartz Cowan so graphically puts it: 'Several million American women cook supper each night in several million separate homes over several million separate stoves' (Cowan 1979, p. 59). It has been impossible to rationalise household production along the lines of industrial production. Domestic technologies that cross the boundaries of the single-family household have been invented, but have persistently failed, even though ownership by individual households is in many cases patently uneconomic in cost terms. And, although these domestic technologies are clearly intended to be used by women, women's interests certainly do not inform the design process. I often wonder how it is that I have such an inefficient cooker and vacuum cleaner when we can fly men to the moon!

The fact is that much domestic technology has anyway not been specifically designed for household use but has its origins in very different spheres. Many appliances were initially developed for commercial, industrial or even military purposes and only later were they adapted for home use. Microwave ovens, for example, are a direct descendant of military radar technology and were developed for food preparation in submarines by the US Navy. They were first introduced to airlines, institutions and commercial premises before manufacturers turned their eyes to the domestic market. For this reason new domestic appliances are not always appropriate to the household work that they are supposed to perform. Nor are they necessarily the implements that would have been developed if the housewife had been considered first or, indeed, if she had had control of the processes of innovation.

Cynthia Cockburn (1992, pp. 32–47) discusses some of the issues which arise when women investigate the appropriateness of such technologies as microwave ovens to activities in the domestic sphere. The relevance of the gender—and by implication the perspective—of the researcher is increasingly recognised by initiatives like the Vienna Centre project which started in 1988. This study brought together women from eleven countries with the intention of stimulating research into the relevance of technological change to wider changes occurring in gender roles and in relationships between the sexes. Typically women's work in the domestic sphere has been invisible to men researching into technology.

There are economic considerations of course—women's domestic labour is unpaid. As an industrial designer I interviewed said: 'why invest heavily in the design of domestic technology when there is no measure of productivity for housework as there is for industrial work?' Instead, when producing for the homes market, manufacturers concentrate on cutting the costs of manufacturing so that they can sell reasonably cheap products. Much of the design effort is put into making appliances look attractive or impressively high-tech in the showroom—for example giving them an unnecessary array of buttons and flashing lights. Far from being designed to accomplish a specific task, some appliances are designed expressly for sale as moderately priced gifts from husband to wife and, in fact, are rarely used. (I'm sure your kitchen cupboards would reveal a range of these devices!) In these ways the inequalities between men and women, and the subordination of the private to the public sphere, are reflected in the very design processes of domestic technology.

Similar forces can be seen at work in the area of contraceptive technologies:

> The newest development in male contraception was unveiled recently at the American Women's Surgical Symposium held at Ann Arbor Medical Center. Dr Sophie Merkin, of the Merkin Clinic, announced the preliminary findings of a study conducted on 763 unsuspecting male students at a large midwest university. In her report, Dr Merkin reported that the new contraceptive—the IPD—was a breakthrough in male contraception. It will be marketed under the trade-name 'Umbrelly'.
>
> The IPD (or intrapenile device) resembles a tiny folded umbrella which is inserted through the head of the penis and pushed into the scrotum with a plunger type instrument. Occasionally there is perforation of the scrotum but this is disregarded since it is now known that the male has few nerve endings in this area of his body. The underside of the umbrella contains a spermicidal jelly, hence the name 'Umbrelly'.
>
> Dr Merkin declared the 'Umbrelly' to be statistically safe for the human male. She reported that of the 763 students tested with the device, only two died of scrotal infection, only twenty experienced swelling of the tissues. Three developed cancer of the testicles, and thirteen were too depressed to have an erection. She stated that common complaints ranged from cramping and bleeding to acute abdominal pain. She emphasised that these symptoms were merely indications that the man's body had not yet adjusted to the device. Hopefully the symptoms would disappear within a year. Dr Merkin and other distinguished members of the Women's College of Surgeons agreed that the benefits far outweighed the risk to any individual man.

This is a parody of a medical research report into the testing of a new contraceptive. The ironic inversion —some may even find it funny—is that here men are the guinea pigs. In real life, contraceptive technologies have been developed by men for use by women. (Interestingly, the incentive for the development of the condom was not birth control but rather men's need for protection from venereal desease.) Despite all the known health risks, for example, the Pill is still one of the most widely used methods of birth control. Its popularity, as against safer barrier methods, points to emphasis upon the sexual pleasure of men and a discounting of the health costs to women.

Technology as masculine culture

Women's absence from the design processes of technology has often been remarked upon. Why is it that so many women feel estranged from, and lack confidence with, technology? In our culture technology is seen as an activity appropriate for men. Different childhood exposure to technology, the prevalence of different role models, different forms of schooling, and the extreme gender segregation of the job market all lead to what Cynthia Cockburn describes as 'the construction of men as strong, manually able and technologically endowed, and women as physically and technically incompetent' (1983, p. 203).

To understand this we need to see technology in terms of not just machines/artefacts, but the physical and mental know-how to make use of these machines. Technological know-how is a resource that gives those who possess it a degree of power, and it is largely possessed by men. Indeed it could be said that appropriating technical expertise is a defining characteristic of masculinity. Men affirm their masculinity through technical competence and posit women, by contrast, as technologically ignorant and incompetent. There is nothing 'natural' about this affinity of men with machines. It has, like gender difference itself, been developed in a social process over a long historical period in conjunction with the growth of hierarchical systems of power. Technology now enters into our sexual identity. Femininity is incompatible with technological competence; to feel technical competence is to feel manly. (See Wajcman 1991, pp. 137–61.)

The example of computing highlights this nexus between masculinity and technology. Here technology is seen as definitive of the activity in question. It is the archetypical case, as to be in command of the very latest technology signifies being involved in directing the future and so is a highly valued and mythologised activity.

This is strongly reflected in Tracy Kidder's (1982) account of a group of men inventing a new computer in *The soul of a new machine*. Here we find a mixture of professional competitive rivalry and complete dedication which characterises the engineers' pursuit of the 'perfect computer'. It is a world of men working compulsively into the small hours, enjoying being stretched to the limits of their capacity, where there is no space for or compromise with life outside of work. It was 'the sexy job' to be a builder of new computers, and you had to be tough and fast; members of the group often talked of doing things 'quick and dirty', and of 'wars', 'shootouts', 'hired guns', and people who 'shot from the hip'. Sexual metaphors abound such that the excitement of working on the latest computer was likened to 'somebody told those guys that they would have seventy-two hours with the girl of their dreams'. It is surely no coincidence that the protagonists of the story are almost exclusively male.

It is evident that men identify with technology and through their identification with technology men form bonds with one another. Women rarely appear in these stories, except as wives at home providing the backdrop against which the men freely pursue their great projects. This masculine workplace culture of passionate virtuosity is typified by the hacker-style work so well described by Sherry Turkle (1984) in a chapter entitled 'Hackers: loving the machine for itself'. Based on ethnographic research at Massachusetts Institute of Technology, Turkle describes the world of computer hackers as the epitome of this male culture of 'mastery, individualism, nonsensuality': 'Though hackers would deny that theirs is a macho culture, the preoccupation with winning and of subjecting oneself to increasingly violent tests [such as working non-stop for days on end] make their world peculiarly male in spirit, peculiarly unfriendly to women' (1984, p. 210).

Being in an intimate relationship with the computer is also a substitute for, and refuge from, the much more uncertain and complex relationships that characterise social life. According to Turkle, these young men have an intense need to master things; their addiction is not to computer programming but to playing with the issue of control. It is about exerting power and domination within the unambiguous world of machinery. These workplace cultures constitute a world from which women are profoundly alienated and from which they are anyway excluded.

Conclusion

In this chapter I have demonstrated the shortcomings of technological determinism. The use/abuse model that represents technology itself as neutral, while its effects are determined by human application, is not an adequate analysis of the relationship between society and technology. Rather, as I have argued, technologies embody power relations. They are always a form of social knowledge, practices and products.

This is so whether it be lightbulbs, refrigerators or machine tools. The same argument can be made in relation to other technologies, such as military, information, or human reproductive technologies. In each case, we need to examine the social context in which the technology is developed. Technological change itself is neither the path to progress nor the road to Armageddon. It is a process subject to struggles for control by different groups, the outcomes of which depend primarily on the distribution of power and resources within society. Perhaps by understanding the dynamics of this process we will be better placed to assert our right to shape it.

References

Bruland, Tine 1985, 'Industrial conflict as a source of technical innovation: the development of the automatic spinning mule', *The social shaping of technology: how the refrigerator got its hum*, eds Donald MacKenzie and Judy Wajcman, Open University Press, Milton Keynes

Cockburn, Cynthia 1983, *Brothers: male dominance and technological change*, Pluto Press, London

——1992, 'The circuit of technology: gender, identity and power', *Consuming technologies: media and information in domestic spaces*, eds R. Silverstone & E. Hirsch, Routledge, London

Cowan, Ruth Schwartz 1979, 'From Virginia Dare to Virginia Slims: women and technology in American life', *Technology and culture*, no. 20, pp. 51–63

——1985, 'How the refrigerator got its hum', *The social shaping of technology: how the refrigerator got its hum*, eds Donald MacKenzie and Judy Wajcman, Open University Press, Milton Keynes

Hughes, Thomas 1985, 'Edison and electric light', *The social shaping of technology: how the refrigerator got its hum*, eds Donald MacKenzie and Judy Wajcman, Open University Press, Milton Keynes

Jones, Barry 1990, *Sleepers, wake!*, Oxford University Press, Melbourne

Kidder, Tracey 1982, *The soul of a new machine*, Penguin, Harmondsworth

Marx, Karl 1976, *Capital*, Vol. 1, Penguin, Harmondsworth

Noble, David 1984, *Forces of production: a social history of industrial automation*, Knopf, New York

Rosenberg, Nathan 1982, *Inside the black box: technology and economics*, Cambridge University Press, Cambridge

Turkle, Sherry 1984, *The second self: computers and the human spirit*, Granada, London

Wajcman, Judy 1991, *Feminism confronts technology*, Allen & Unwin, Sydney

Annotated bibliography

Cockburn, Cynthia 1983, *Brothers: male dominance and technological change*, Pluto Press, London
A classic study of the effects of technological change upon the organisation and experience of paid work for a group of archetypal male craft workers.

Cowan, Ruth Schwartz 1989, *More work for mother: the ironies of household technology from the open hearth to the microwave*, Free Association Books, London
An excellent, readable history of housework in America focusing on the relationship between technological development and household work.

Kirkup, G. and Keller, L. S. 1992, *Inventing women: science, technology and gender*, Polity Press, Cambridge
Exploring the gendering of science and technology, this edited volume looks at women as producers and consumers.

MacKenzie, Donald and Wajcman, Judy 1985, *The social shaping of technology: how the refrigerator got its hum*, Open University Press, Milton Keynes
Now a standard reference collection on the social construction of technology; especially the technology of production, domestic technology and military technology.

Probert, B. and Wilson, B. 1993, *Pink collar blues: work, gender and technology*, Melbourne University Press, Melbourne
This Australian collection looks at gender as a key dimension of power relations in the workplace, raising issues about women's skills, pay and work organisation.

Wajcman, Judy 1991, *Feminism confronts technology*, Allen & Unwin, Sydney
Provides a comprehensive and accessible review of feminist theories and research on technology, including chapters on workplace, domestic and reproductive technologies; and technology as masculine culture.

2

VIRTUAL REALITY FAKES THE FUTURE: CYBERSEX, LIES AND COMPUTER GAMES

David McKie

In essence virtual reality (VR) is simple. Its potential applications are extraordinary; its implications immense. Simply defined by the phrase, 'interactive graphical simulations', it revitalises the eternal cyberquestion, 'can human be merged with non-human and/or with machine? Reality is so restrictive—is it possible to move beyond it into an infinity of potential cyberworlds?

Current understandings of VR, by overemphasising technological equipment, limit considerations of VR to the narrow frame of human perception of computer-generated worlds which are experienced and interacted with via sensors. The experiential/perceptual dimension makes the crucial distinction from intellectual interaction with mere multimedia. Initiates insist that if you want to understand VR then you have to wear a computer and feel the difference.

Descriptions of the necessary sartorial hardware occupy much VR literature. Head Mounted Displays (HMDs in trade jargon) resembling Darth Vadar helmets include a video camera pathway over each eye to allow visual exploration of an artificial environment—see Marilyn Monroe's pout in hair-follicle close-up. VR data gloves, which act as gesture recognition devices, unlock different doors to the same kind of virtual experience and permit virtual interactions with virtual three-dimensional objects. Wanting more? In some advanced systems whole-body datasuits, wired with position and motion transducers, allow users to transmit to others, and/or to represent to themselves, the shape and activity of their body in the virtual world.

Cybersex, cyberlies and exaggerations

Short of actual hand-in-glove, helmet-on-head or body-in-suit experience, illustrations help to fill out the current meaning spectrum of VR. Significantly, the word-of-mouth descriptions, such as the lists of applications included in Howard Rheingold's (1992) influential book *Virtual reality*, cluster around heavy technology. Even without VR's potential future, current interactive graphical simulations are revolutionising certain existing technological practices. Rheingold describes how:

> I commanded repair robots in virtual 'outer space' [at NASA] . . . I ran my fingertips over 'virtual sandpaper' by means of a texture-sensing joystick and watched scientists create animated creatures who will live in tomorrow's semi-sentient virtual worlds [at Cambridge] . . . I played with . . . a pair of goggles and gloves that take you—through the eyes and hands of a robot—under the ocean or into your own bloodstream [in Vancouver] (Rheingold 1992, p. 18).

Long before Rheingold reaches Silicon Valley and dances with a woman in 'the form of a twelve-foot-tall three-dimensional purple monster', the cumulative extraordinariness of the descriptions makes it easy to understand why public attention has been gripped by VR's high-tech dimension. Such experiences, fresh, sensational and visually rich, come ready made for newspaper headlines and spectacular media special effects.

Then, of course, there's the future promised by prototypes—especially in cybersex or 'teledildonics'. (From 'dildonics', allegedly used to describe a machine that converts sound into tactile sensations—although dildo is more commonly used to refer to an artificial penis—and 'tele' to describe communication over a distance.) With teledildonics some people may soon—supposedly in a silicone bush, glen or valley not too far from you—be able to wriggle into a condom-tight bodysuit embedded with thousands of miniature electronic sensors, computer controlled to simulate the feel of any object from rubber to skin. Suitably protected participants could then sexually interface, in an AIDS-free, body and mind blowing experience, with some other individual or group—real, imaginary or re-created. The only compatibility requirement for a virtual orgy would be network access and the desire to log in.

Given such uses and potential gratifications it is little wonder that forecast applications along these lines have captured the main public sense of VR as an interactive s[t]imulation mainly concerned with touchy-feely sensoramas. Where the public imagination has been stretched in other directions, it has been towards more frightening arcade games and bigger toys (smart bombs) for big boys. The discourse is masculine and neurotically concerned to avoid consideration of emotional implications, virtual or otherwise.

Exaggerated predictions based on teledildonic fantasies also fuel unrealistic expectations of immediate technological breakthroughs. Designs for amazing computer-aided future functions abound, but so far the exciting possibilities appear to be a long way from realisation. As one science journalist remarks—commenting upon the VR hype for 'impending' hyperrealities—'Real estate is cheap, but the places you'd most want to visit are still under construction' (Antonoff 1993, p. 83).

Undeterred by the gap between VR realisation and public expectations, advocates such as the *CyberEdge Journal* editor declare confidently that VR is where personal computers were in 1979: 'PCs back then were slow . . . But you could start to see the promise. Ten years later everything was changed. Virtual reality may have a little longer gestation period, but it has the same potential' (quoted in Antonoff 1993, p. 125). Rheingold's book cover lists other less measured claims: VR 'won't merely replace TV' but 'eat it alive'; VR 'may be the most important development since man first chipped flint' and VR 'will represent the greatest event in human evolution'.

Apart from selling books, the underpinning agenda here is familiar. Publicity helps to attract funding and makes 'cyberlies for megagrants' more likely. With sufficient money invested, eventual quantum leap breakthroughs are virtually guaranteed. Meanwhile, more likely to be featured on *Quantum* programs than in *Playboy*, useful and down to earth applications gain limited recognition. Yet some of these—medical imaging for example—begin to hint at specific, and closer to realisation, utilitarian potentials of interactive graphical simulations. In the words of legendary VR practitioner, Professor Fred Brooks: 'The technology advances best if you *carefully* choose a good driving problem, with good collaborators who will keep you honest and keep your feet on the ground' (quoted in Rheingold 1992, p. 39).

Brooks' selected co-workers at the University of North Carolina come from a list which includes astronomers, geologists, chemists, architects and highway safety people. If the future is going to be virtually faked in advance, then his advice about relative honesty, and intelligent collaborators with actual problems, shortens the odds in favour of development in a socially useful direction.

As an example of responsible research, Brooks' team initially focused on haptic perception ('that melange of senses we lump together under the category of "touch" . . . and the body's internal sense of *proprioception* that informs us about the position of our own limbs in relation to one another and to the space around us'), combining that understanding with computers to create tools for biochemists working with protein and nucleic acid structures (Rheingold 1992, p.27).

Contradictions in cyberspace: from here to hypereternity

Brooks' North Carolina campus also offers experience of virtual architecture and versions of cyberspace. At the simplest level the university's virtual cooks can ergonomically test layouts with virtual pots in the campus virtual kitchen. On a larger scale virtual visitors can enter virtual campus buildings for a sense of what it feels, looks and sounds like for them to walk through, or to live in, such places. Elsewhere in the world recently imaged buildings include a virtually reconstituted French cathedral (which actually collapsed in 1815). The advantage for architects and potential property buyers, never mind medieval scholars, who can use VR images prior to surveying a site, is an obvious and clearly practical use of the technology. These cyberspaces can be defined as 'an infinite artificial world where humans navigate in information based, and computer generated space'—dovetailing virtual architecture neatly with conventional VR meanings and applications.

At the other extreme various thinkers conjure up more mystical definitions. If technological VR is too restrictive then mythological cyberspace runs the risk of being too expansive. William Gibson, whose novel *Neuromancer* first used the term in 1984, introduced cyberspace as

> consensual hallucination . . . graphic representation of data abstracted from the banks of every computer in the human system. Unthinkable complexity. Lines of light ranged in the nonspace of the mind, clusters and constellations of data (Gibson 1984, p. 51).

Through the largely non-fictional mode of an essay collection, Michael Benedikt (1992), professor in the School of Architecture at the University of Texas, attempts to merge the two extremes. Bathed in the afterglow of the fall-out from failed predictions of the paperless office and the cashless bank, one of Benedikt's many definitions of cyberspace displays a sense of mystery and an unusual environmental awareness:

> The realm of pure information, filling like a lake, siphoning the jangle of messages transfiguring the physical world, decontaminating the natural and urban landscapes, redeeming them, saving them from the chain-dragging bulldozers of the paper industry, from the diesel smoke of courier and post office trucks, from jet fuel fumes and clogged airports, from billboards, trashy and pretentious architecture, hour-long freeway commutes, ticket lines, and choked subways . . . from all the inefficiencies, pollutions (chemical and informational), and corruptions attendant to the process of moving information attached to *things*—from paper to brains—across, over, and under the vast and bumpy surface of the earth rather than letting it fly free in the soft hail of electrons that is cyberspace (Benedikt 1992, p. 3).

After that quote it is not surprising that Benedikt's collection, *Cyberspace: first steps*, includes a speculative short story by Gibson. Alongside it, however, are more down to earth definitions, and prosaic applications such as corporate virtual workspaces. The cyberspace concepts activated in the essays—contributed by architects, scientists and software engineers—would make for an impossibly fat dictionary, but they certainly widen ideas of cyberspatial scope. In addition they throw up a composite definition which, while it might still describe an electronic orgy, allows mystic and electronic meanings to peacefully coexist in a 'democratised' forum:

Cyberspace is a completely spatialized visualization of all information in global information processing systems, along pathways provided by present and future communications networks, enabling full copresence and interaction of multiple users, allowing input and output from and to the full human sensorium, permitting simulations of real and virtual realities, remote data collection and control through telepresence, and total integration and intercommunication with a full range of intelligent products and environments in real space. (Novak, in Benedikt 1992, p. 225)

The democracy implied in 'full copresence' serves as a useful reminder of the widespread nonparticipation of the vast majority of the world's population in both VR and cyberspace. For, despite the inclusive rhetoric of future potentials for all, traditional economic, race and gender lines mark the usual restricted points of access. Whether speaking as engineering technocrats or cyberspace visionaries, leading players remain white, male and middle class. They devote more attention to constructing retinal imaging lenses for mind sex, military hardware or more virtually real computer games than to the promotion of wider participation.

Cyberpunk fiction acknowledges that deprivation coexists with cyberspace and imagines nonvirtual living conditions for the unnetworked, beyond the hallucinatory 'matrix' of globally interlinked terminals. By putting the punk, in the sense of the 'socially unimportant person', back in the cyber, cyberpunk gives its cyberspace greater social relevance than VR. Visible all along Gibson's 'sprawl' (bleak built-up strips linking previously separate cities and mega-metropolitan centres), urban decay spawns 'lo-teks' (low technique, low technology), 'moderns' (nihilistic technofetishist) and other subcultures of self- and socially-mutilated young people. Outside the information and technology rich network dominated—not that futuristically—by near omniscient corporate power and multinational underworld groupings, most cyberpunks live lives of dystopic desperation.

Reconstituting virtuality: from *Jurassic park* to the new biorevolution

One radical feature of cyberpunk remains unexplored by VR: the mutation of the organic building blocks of living creatures—the skin, bones, sinews and senses—to incorporate the cyber. Gibson's characters undergo various kinds of bioengineered hi-tech graftings ('toothbud transplants from Dobermans'), cranial data implants or electronic prosthetics, but stop short of mutation. Yet the possibility of mutation marks the return of the cyberquestion—can human be merged with non human and/or with machine? Hybrid cyberorganisms (cyborgs)—usually played by Arnold Schwarzennegger—have revitalised Hollywood blockbusters and drawn both public attention and cultural comment. In films such as *Bladerunner*, *Robocop*, *Robocop 2*, *Terminator 2: judgment day*, and *Total recall*, the cyborgs, replicants and other alien forms not only contain organic matter but exhibit more humanity than the humans. . .

Hollywood's fascination with the cyborg has become laminated with an increasing use of morphing techniques which can realistically turn an onscreen Michael Jackson into a lifelike panther—and take enough age from Clint Eastwood to make him a credibly young bodyguard at John Kennedy's assassination. Director James (*Terminator*) Cameron observes that seven out of the ten all-time top grossing films have been made with hugely sophisticated visual effects.

Jurassic park rather than *Lawnmower man* sets the benchmark for VR at Hollywood. It also foregrounds narratively the unpredictable organic component in virtual dinosaurs and, technologically, the stunning simulations of digital hyperrealism—the realism of the dinosaurs contrasts favourably with the acting of the humans. But verisimilitude need not stop with the animals. In one scene, where a raptor leaps at the ceiling to capture the young girl, Industrial Light and Magic technicians gave the impression of the actual child actress being present by overlaying her face on top of the body of the stunt woman who actually hung in that dangerous position. One leading special effects man has subsequently commented in a radio interview that

> people are often afraid of the possibility of now having digital human beings replace real human beings. From my point of view that may be a possibility but it's immaterial, all we're interested in doing is making pictures and telling stories and when a person will suffice or be the best way to do it that's what we'll use (Dippé 1993).

Actors' Equity is unlikely to be reassured.

More importantly for non-actors, how far will life mimic popular art? San Diego Zoo's Center for the reproduction of endangered species has projects which include a contemporary ark to maintain frozen genetic material from some 350 endangered species. The human genome diversity project made plans to collect blood from individual members of ethnic groups facing extinction by using recently available methods of gathering and saving human cells and identifying DNA variations. While not intending to recreate them for a theme park, the project attracted strong criticism for applying late twentieth-century technology to nineteenth-century racist biology, and for allocating millions to blood sampling while the indigenous peoples involved struggle to maintain their lives and cultures (Lewin 1993a, p. 25).

For some time biotechnology has suffered the same 'cyberlies for megagrants' overhype as VR; and its pigs still can't fly. Recently, however, it has regained respectability—attracting funding for smaller and more achievable applications. During the past decade, 1300 or so new biotechnology companies with a combined annual turnover of around $US8.1 billion, have emerged (Coghlan 1993, pp. 26–7). The exploitation of 'homologous recombination' (a way of inserting DNA at will into a chromosome), by such companies as Cell Genesys in America, raises more fundamental technology/nature issues than do cyborg movies and teledildonic dolls.

Analysis of the techno-biblio-fundamentalist names in the biotechnology industry indicates cause for concern: Genentech, Genzyme, Cell Genesys and Gilead (shades of Atwood's future world in *The handmaid's tale*). Spurred on by profit, not social responsibility, the name of the game is creation. The promotion slogan for *Lawnmower man*, 'God made him simple, technology made him a god', and the comic VR question, 'did reality move for you?', take on a whole new dimension when biotechnology companies are designing shifts in genetic makeup. The degree of virtuality when technical components merge with this kind of organic material warrants more critical attention. Where does the virtual become real?

In Princeton experimental transgenetic swine are being bred with human genes to develop suitable haemoglobin for blood transfusions. This program, and others like it, have arisen to meet the market in alleviating human panic over the possibility of AIDS infected blood products. Pigs still lack wing buds, but with genetic augmentation they look likely to bleed to profitable effect in an expensive propitiation of an irrational human fear. Beyond the USA an Edinburgh firm has genetically engineered livestock capable of producing valuable pharmaceuticals in milk. Although Rheingold speculates briefly on how future VR convergence with nanotechnology may create 'microscopic robots capable of navigating through blood vessels' (Rheingold 1992, p. 362), VR's restrictive emphasis on white, male entertainment technologies contributes to the comparative neglect, and low public awareness, of these other crucial virtualities already in production.

Reframing future agents (1): tales from the internet

To return to more conventional VR—cyberspace and its intersection with identities—I propose to look through the internet at questions of access in relation to prospects for indigenous peoples and women. First a typical high publicity male puff, from an article by Robert Wright, republished in *The Australian*. How virtually honest is the given real in networked virtual reality?

Wright makes significant claims for his understanding of internet—that vast and diffuse 'network of networks' spreading its electronic web worldwide—as 'an expression of human nature by the most efficient available technology'. After a few weeks of logging in, he confidently concludes that:

> no Netregulars use true anonymity to be purposefully abusive . . . [and as] for the sexually liberating effect . . . where inhibitions dissolve, genders get switched, fantasy roles played . . . this is an infinitesimal drop in the bucket, irrelevant to what 99.9 per cent of all internauts are doing (Wright 1993, p. 33).

Wright's conclusions look less credible when set against Allucquere Rosanne Stone's account of an actual case. The main participant, whom Stone calls 'Julie', logged into the net in the persona of a totally disabled older woman only able to push computer keys with a headstick: 'Her standard greeting was a big expansive "HI!!!!!!". Her heart was as big as her greeting'. In the intimate electronic companionships that subsequently developed, 'Julie's women friends shared their deepest troubles, and she offered them advice—advice that changed their lives'. Yet 'Julie' later turned out to be an able bodied middle-aged male psychiatrist who had constructed an artificial net persona to gain access to women's net talk which had 'so much more vulnerability, so much more depth and complexity' than men's. After someone discovered his real identity, the fall-out 'varied from humorous resignation to blind rage' with one woman feeling raped because her 'deepest secrets had been violated', and with others repudiating 'gains made in their personal and emotional lives' because they had been 'predicated on deceit and trickery' (Stone, in Benedikt 1992, pp. 82–3).

Clearly Wright's simplistic assurances do not meet the fluid possibilities of extended time on the net. Ironically, while allowing the existence of deception, Stone's tale contradicts the pessimism about identity blurring by illustrating how the internet also permits gender- and disability-disregarding possibilities. For some—who may have a real persona at odds with their virtually real physical bodies—the internet permits actualisation of their different self. At the same time Stone shows how deeper female ways of relating can turn the net to good emotional use.

The potential for emotionally-deep women's networking supports what Sherman and Judkins (1992) discern as 'shreds of evidence' that VR might offer 'a new world for women'. As the binary divide of the title—*Glimpses of heaven, visions of hell*—signals, Sherman and Judkins more than counterbalance their optimism with pessimistic details of declining female participation rates and substantial Pentagon economic involvement in VR research. Nevertheless there is a heartening logic to their positive view that the over-technical nature of current computer interfacing discourages many women and that, with more natural VR interfacing, 'which militates against their specialised faculties, men will lose their confidence initiative' (Sherman & Judkins 1992, p. 165).

More inspirationally Donna Haraway argues that women are already cyborgs in the sense of being simultaneously human and technological. As testament Haraway offers her own historical position:

> a PhD in biology for an Irish Catholic girl was made possible by Sputnik's impact on US national science-education policy. I have a body and mind as much constructed by the post-Second World War arms race and cold war as by the women's movements (Haraway 1991, p. 173).

Undoubtedly though, to make the most of current VR, you need to be male, speak English and be wealthy (or, at least, able to access the facilities of industrialised nations). Currently over half the world's population cannot log in to that most essential bottom line—the phone network. Database availability is similarly skewed in favour of the already information rich. Many well-stocked Western libraries face underuse because researchers based there obtain their information needs electronically. Meanwhile users of under-resourced third world libraries—which lack good collections and have most to gain from database access—are often unable to afford basic connection costs (Holderness 1993, p. 37).

As with women and technology, the fact that this imbalance is largely so need not mean that it will always be so. The well known work of Eric Michaels, *For a cultural future: Francis Jupurrurla makes TV at Yuendumu* (1987), shows an Aboriginal interaction with television which disturbs the usual unidirectional flow of media control and innovation, offering an image of a democratised technology. In a recent special issue of *Media information Australia—Art & cyberculture*, other theorists have argued congruent cases for the technology as highly appropriate to the lives of indigenous people. This is especially true where the traditional culture virtually prefigures the networks of the industrial world—with the important proviso that all communications are grounded in appropriate psycho-social understandings. David Tafler and Peter d'Agostino (1993), for example, suggest how 'the formulation and exchange of stories' from indigenous peoples, can 'recuperate technological advances that have distanced human societies from the rigours of their environment':

> Bridging generations, stories tell of forgotten experiences while transmitting knowledge, lineage and insight. Insofar as they transcend spatial and temporal constraints, the formulation and exchange of stories represent the harbinger of more advanced telecommunications (Tafler & d'Agostino 1993, pp. 51–2).

Such insights suggest the possibility of a more balanced cultural exchange between indigenous people and urban populations. Already, when Native American VR warriors go cyberspacing as networked nations, they employ software in a vibrant fusion of the oral and electronic:

in the hands of the Sioux, Crow, Navajo and Assiniboine artists, NAPLPS [the network system] has a powerful new identity: a runner crossing snow-capped mountain tops, a messenger flying across the land of the Crow on the back of the big-beaked bird, telling stories to the children of the Sioux, blowing magic into the eyes of Assiniboine children, bringing ancient messages into the heart of the future (Hedlund 1992, p. 32).

Reframing future agents (2): cyborg subjects in cybernature

My final attempt at framing the VR future, or, more accurately, reframing the parameters of what is to count as VR in the future, starts with a passage by Norie Neumark:

> We are living in a moment of computer-related crisis. We fashion ourselves as computers. We feel hardwired. We scan our memory banks and databases to access information. Gone are the days of the mechanical self with plumbing and tubes, sparking on all four cylinders . . . now many of us think of ourselves as microcomputers on legs (Neumark 1993, p. 80).

Disagreeing profoundly with her, I use Neumark's quote (which doesn't reflect the overall subtlety and stimulation of the article) for two reasons. Firstly, the passage catches the technological overkill so destructive in VR's faking of the future, while simultaneously revealing its Achilles heel. Secondly, it leads into what I see as the most exciting virtuality field: complexity—the emerging microcomputer science at the edge of chaos theory.

My disagreement with Neumark stems from her too easy dismissal of the actual body organism and the too quick leap into the cybertech half of the cyborg hybrid. Pretending that body matters don't matter is an old Cartesian mind/brain trip that technocrats regularly fall over, and which remains liable to cause VR pratfalls or worse. Presently recovering from a virus I feel nothing like a microcomputer on legs, but it is very easy to forget such things when the body is well and the technology is powerful and new.

Computers now are not so new but what their augmenting power might do, and what their paradigm shifting potential might be, is still being explored. Rheingold's account of VR opens with a 1988 quote speculating that through the computer we 'may begin to see reality differently' simply because it 'produces knowledge differently from the traditional analytic instruments' and 'provides a different angle' (Pagels, quoted in Rheingold 1992 p. 15). By quarantining themselves off as a distinctively technological field I believe VR advocates miss the potential for transformation using simpler, less spectacularly hardwared, paradigm shifts. By contrast, the field of complexity builds on seeing reality differently by exploring computer generated 'life' through interactive graphical simulations.

Working with the idea that a few simple rules lie at the root of all complex systems, complexity work holds out the promise of a unifying theory of biological, economic and physical phenomena. Probably the most famous, and influential, instance still remains James Lovelock's 'Daisyworld' software—a small-scale self-generating near embodiment of a digital universe (Lewin 1993b, pp. 108–19). That simple program already impacts significantly on perceptions of what the earth is, revises conceptions of humanity's place on it and suggests the planet is so virtually real that it is as alive as a person.

Also—without data gloves or HMDs—complexity theorists have run self-generating models of everything from stock markets to the rise and fall of ancient civilisations. Central to the complexity concept are notions of natural self-organisation, whether it be bacteria, pre-biotic life or industrial companies. This framework sets up new relations between laws of nature and patterns of behaviour. In evolutionary biology, for example, software can self-generate systems remarkably similar to actual species evolution.

In the course of his book *Complexity: the emerging science at the edge of chaos,* M. Mitchell Waldrop (1993) discusses the work of Chris Langton (whom he terms 'the architect of artificial life'). Langton studies the virtual lives, and virtual behaviour, of cellular automata created algorithmically in a computer. Both Langton and his research mark a significant step in virtuality—unless, perhaps, the life created is real? Investigations of what currently counts as VR tend, as Rheingold's (1992) book does, to ignore this field. It is a significant omission. Why concentrate upon cybersex without considering the possibility of faking the organism?

Science aspires to the so-called mind of God because scientific equations appear to match what happens in the universe. Complexity's virtual realities apparently mimic the spontaneous ordering of natural life forms. Complexity's interactive graphic simulations seem capable of generating natural crystallising patterns from randomness. As with the further boundaries of VR's faked futures the jury is still out. (Complexity theorists too might just be playing more computer games.) Yet my guess is that complexity's low-tech imagination will leap ahead in creating virtually real behaviour in virtually real worlds. Complexity, rather than high tech VR, may be the real future currently being faked. What does that mean for reframing human identity?

In the final confessions of a virtual realist, I favour taking a less anthropocentric perspective of our place as humans in the universe. Animal rights activists, environmentalists, ecofeminists and deep ecologists have been arguing that position from an organic perspective. Even the architect of artificial life, Langton, admits that he 'can't think of evolution acting on individuals any more . . . It's always acting on an eco-system, a population, with one part producing something another part needs' (quoted in Waldrop 1992, p. 359). Other technosciences, for example VR bioengineering, support the likelihood of a paradigm shift by promising equally radical upheavals in that disciplinary perception of who we are and what makes up life. The move was prefigured discursively back in 1985 with Donna Haraway's conception of dialectical interplay:

> Cyborg imagery can suggest a way out of the maze of dualisms in which we have explained our bodies and our tools to ourselves. This is a dream not of a common language, but of a powerful infidel heteroglossia. It is an imagination of a feminist speaking in tongues to strike fear into the circuits of the super-savers of the new right. It means both building and destroying machines, identities, categories, relationships, space stories. Though both are bound in the spiral dance, I would rather be a cyborg than a goddess (Haraway 1991, p. 181).

Conclusion

In faking the future, VR would do well to remember that environmental reality is not virtual and equity issues must be addressed. Not everyone on the globe has access to food—let alone the internet. Dreaming the impossible dream, technologically predicated on 5.3 billion microcomputers, arcade games and terminal points, VR posits an impossible individuality incompatible with planetary survival.

Alongside feminist projections of cyborgs, are other tentative gropings towards cyberspace, cyberpunk and complexity. Without donning HMD helmets, and prior to direct retinal imaging, these forward-thinking groups and individuals acknowledge the possibilities of virtual realities without cybersex, lies or computer games. In valuing our biological heritage, they focus their discursive strategies on the need for collective local and global action. Avoiding the hype of hi-tech hyper-reality, they offer a more responsible, more playful way to create myths about the future, while rendering the present more precious and more real.

References

Antonoff, M. 1993, 'Living in a virtual world', *Popular science*, June, pp. 82–6 & pp. 124–5

Benedikt, M. 1992, *Cyberspace: first steps*, MIT Press, Cambridge, Mass.

Coghlan, A. 1993, 'Engineering the therapies of tomorrow', *New scientist*, 24 April, pp. 26–31

Dippé, M. 1993, *Screen* (Radio national), 9 September, interview

Gibson, W. 1984, *Neuromancer*, Grafton, London

Haraway, D. J. 1991, *Simians, cyborgs, and women: the reinvention of nature*, Routledge, New York

Hedlund, P. 1992, 'Virtual reality warriors: Native American culture in cyberspace', *High performance*, spring, pp. 31–5

Holderness, M. 1993, 'Down and out in the global village', *New scientist*, 8 May, pp. 36–9

Lewin, R. 1993a, 'Genes from a disappearing world', *New scientist*, 29 May, pp. 25–9

——1993b, *Complexity: life at the edge of chaos*, Orion Publishing, London

Michaels, E. 1987, *For a cultural future: Francis Jupurrurla makes TV at Yuendumu*, Artspace, Melbourne

Neumark, N. 1992, 'Diagnosing the computer user: addicted, infected or technophiliac', *Media information Australia*, no. 69, August, pp. 80–7

Novak, M. 1992, 'Liquid architecture in cyberspace', *Cyberspace: first steps*, ed. M. Benedikt, MIT Press, Cambridge, Mass. pp. 225–72

Rheingold, H. 1992, *Virtual reality*, Mandarin, London

Sherman, B. and Judkins, P. 1992, *Glimpses of heaven, visions of hell: virtual reality and its implications*, Hodder & Stoughton, London

Stone, A. R. S. 1992, 'Will the real body please stand up?: boundary stories about virtual cultures', *Cyberspace: first steps*, ed. M. Benedikt, MIT Press, Cambridge, Mass. pp. 81–118

Tafler, D. and d'Agostino, P. 1993, 'The techno/cultural interface', *Media information Australia*, no. 69, August, pp. 47–54

Waldrop, M.M. 1992, *Complexity: the emerging science at the edge of chaos*, Viking, London

Wright, R. 1993, 'Meetings of ghostly minds in the machine', *The Australian*, 8 September, pp. 22–3

Annotated bibliography

Gibson, W. 1993, *Virtual Light*, Bantam, New York.
His new (see Michael Galvin's bibliography entry for important earlier work) post-cyberpunk novel which, despite its title, says little about VR directly but implies much by how it treats the privatisation of public space.

Lewin, R. 1993b, *Complexity: life at the edge of chaos*, Orion Publishing, London
An honest attempt to come to terms with a difficult field personally, as well as introducing it to his readers and evaluating its claims to be the successor to chaos. Informed by his own science background it is usefully wide and, unlike Waldrop's book on the same subject, not afraid to be critical.

Rheingold, H. 1992, *Virtual reality*, Mandarin, London
Still undoubtedly *the* place to start with VR. Rheingold tried to experience

everything, talk to everyone and set the whole in a stimulating, and helpfully documented, context.

Sherman, B. and Judkins, P. 1992, *Glimpses of heaven, visions of hell: virtual reality and its implications*, Hodder & Stoughton, London
Accessible, issue-focused introduction to the field with many stimulating analogies despite a sometimes irritating either/or approach. Its helpfulness is hampered by the total absence of references and bibliography but it is still very readable.

Woolley, B. 1992, *Virtual worlds: a journey in hype and hyperreality*, Blackwell, Oxford
A rather odd essay–journalist approach not nearly so black and white as the Sherman and Judkins book nor as dismissive as its own title suggests. More wide ranging and eclectic than other VR literature and well worth reading for that alone.

THE TECHNOLOGY OF TELEVISION
Albert Moran

The British historian Asa Briggs (Briggs 1960) has noted a parallel between a range of discoveries and inventions at the end of the eighteenth and the nineteenth century. Those in the eighteenth had to do with chemical discoveries and mechanical inventions which gave rise, early in the new century, to what has been collectively called the Industrial Revolution. So too a range of discoveries and inventions at the end of the nineteenth century—which included the development of the motion picture, the gramophone and the isolation of the principle of radiated electrical transfer—has in the twentieth century given rise to what has inadequately been described as the Communications Revolution.

Cinema and broadcasting are only a few years off celebrating their centenary birthdays and over the course of their existence there has developed a complex series of interactions and interrelations between these and other technologies which Trevor Barr sees as constitutive of *The electronic estate* (Barr 1985). The technologies are integrated into a society which itself has been massively transformed and made much more complex than a century earlier. Where time and geography once served to delimit societies and nations, one from the other, now it is the technologies of entertainment, communication and information (ECI) that help maintain national cohesion and interrelations.

The family of ECI technologies

Until the development of the electric telegraph in the late 1830s, communications and transport were one and the same phenomenon. A letter could move from sender to receiver only as fast as living or mechanical transport could carry it. Although communication and transport are no longer as intimately linked as they once were, nevertheless the combined effects of these two motive systems continues to have a profound effect on the social, political, economic and cultural fabric of human society. Indeed recent analyses (Anderson 1992) of the profound shifts in the present world order see the decay of national political entities and the outbreak of more regional and local nationalisms as complementary forces driving the change alongside the development of the car and aeroplane.

A particularly vivid illustration of the conjunction of transport and communications occurred in 1969 with the Americans landing on the moon. Such an event had for nearly a century been projected imaginatively in novels, paintings, comics and newspapers so that when the first step on the moon's surface finally came to pass, it was banal and mundane. However, what earlier speculation had entirely missed was the possibility that the event would be simultaneously available to a television audience even as it happened, instead of being something that was reported days or weeks after it took place. Thanks to a simultaneous television and satellite hook-up, the moon landing was the first truly global broadcast experience in the history of earth, seen by more people than any other event in history.

Although it is not quite the most important technology to emerge in the twentieth century, television is certainly one of two or three such inventions (Watson 1990; Barnouw 1978). Yet television was never an autonomous technology which once invented by Baird and others in the 1920s dictated the terms of its own incorporation as an institution. Television, as a technology of communication, is part of a large family of media which, as the following diagram suggests, stretches back to the fifteenth century invention of the printing press (Hadley, McNulty & Salter 1979). Nor is the family tree complete. It is growing at an exponential rate and we might also mention a series of related technologies. A more complete listing of family members would include the following: mobile phones, car phones, beepers, answering machines, cable television, colour and high-definition television, the video recorder, the videotape cassette, remote (radio) control, the video phone, stereophony, video games, electronic reproduction facilities, electronic high-speed printing, microfiche with electronic access, facsimile, electronic mail, printing by radio, compact discs and digital audiotape.

Of course these technologies are not all of the same kind and it is useful to describe them as the ECI clan. Although there are all kinds of overlaps and connections, nevertheless there are important differences. For example, television in the west (Morley 1991) is centrally about entertainment (even if it ritually seems to be about the other two functions, what it does offer in the way of information and communication is mythic).

The notion of entertainment has its own legacy going back to the eighteenth century. In contradiction to an older form of thinking which saw the purpose of art (including popular art) as moral tract, entertainment was a more secular notion that emphasised an art devoid of serious moral or didactic purpose, and given over instead to allowing the audience a pleasurable, even escapist, engagement with the opera, play or novel. Entertainment has extended its range of meanings so that as it has become synonymous with such centres of show business as Tin Pan Alley, Broadway and Hollywood. It implies a notion of spectacle, diversion, escape; giving the people what they want; the mysterious processes and distribution of star

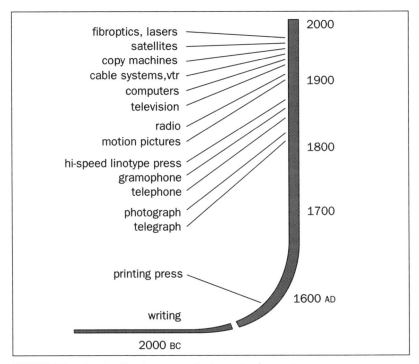

FIGURE 3.1 Acceleration in communications development

quality and so on (Dyer 1978; Dietrich Fischer 1979). Home entertainment may have begun with modest domestic toys such as the piano, the lantern slide show, and the gramophone but in the twentieth century it has spread to embrace both the cinema and radio.

Cinema here is the odd relation. Developed at the end of the nineteenth century, the cinema is a technology of public rather than domestic entertainment. Yet it has been of crucial importance in the subsequent development of broadcast, both radio and television. In the first and second decades of the twentieth century, the cinema (especially the American cinema) developed the main genres of content, the principle of intelligible organisation of sound and images, the production organisation and the marketing and distribution skills that would later be embraced by television and radio (Bordello, Thompson & Steiger 1985). Most especially television has gone beyond film in being home cinema, while it has gone beyond wireless broadcasting in being radio with pictures.

However, rather than seeing television as being adaptive and parasitic in terms of these earlier technologies of entertainment, the crucial tendency in the field of entertainment technologies is towards multiplication and convergence. This comes about because television, radio, cinema, recorded music and so on—the technologies of entertainment—are more than just technologies. The new definitions of what the technologies can achieve in domestic settings, the new institutional arrangements and relations across these media, come about because the technologies in question are social, economic, political and cultural even as they are technical (Kitross & Sterling 1980).

Thus the institutional changes that came over the fields of both broadcast radio and the cinema in the face of the advent of television came about not because there had been any crucial changes in the technology of radio or cinema but because television usurped many of the central practices, contents and audiences that radio and cinema had held in the past (Czitrom 1982). Radio and cinema were forced to redefine their place in the arena of television and they did this finally by seeing themselves as complementary to television.

Similarly in the 1990s with the onset of pay television, new expanded delivery systems, and the general moves to both the expansion and deregulation of video, the era of dominance by broadcast television is coming to an end. The institutional field of television is changing so that broadcast television will in the future be only one way of defining the technology. In other words, the apparatus has never been fundamental in determining how television could be incorporated into society. The technology did not dictate the terms of its incorporation. It only provided a set of possibilities. Instead it was a long chain of human choices, customs, practices and modes of thinking across a diverse series of fields including the economic, the ideological, the social and the political which caused television to be the way it is (Boddy 1979; Curran & Seaton 1989).

In the 1960s the coverage of the Kennedy assassination and then the televising of the moon landing heralded the advent of world television. It also brought a theorist, Marshall McLuhan, to public attention. He was himself partly constructed by the media as a media celebrity. McLuhan had begun his theoretical contribution in the 1940s with work culminating in the 1950s with *The mechanical bride* (1967) which, in its pessimistic reading of the impact of mass communications on western society, was a distinct echo of other cultural conservatives such as F.R. Leavis and Theodore Adorno.

By 1964 his ideas had come a full circle. In *Understanding media*, especially Chapter 1, 'The medium is the message', McLuhan's theory of media is a celebratory one of a technology that has its own ethos outside human society and culture. History is outside human agency and control and is in the hands of the media. McLuhan's value was to emphasise the importance of media as a technology in the development of human society, yet the emphasis on technology was finally so relentlessly deterministic and circular that McLuhan's theories have largely been discounted. If McLuhan's ideas have seemed of less importance in the last 20 years, it is because of the rise of a more general theory of contextualism in the humanities. Raymond Williams, the British Marxist intellectual, has been the most influential figure in shaping the understanding of television and other media for several generations of English and American scholars. Williams' concern was with the broad field of culture in the anthropological sense. Like McLuhan he began as a literary scholar and like McLuhan he broadened his concerns to include the media. However, unlike McLuhan, he never fell into a mechanistic determinism. His short seminal reflection on television (Williams 1974) had the subtitle: *technology and cultural form*, and the bracketing of the two underlies Williams' concern to emphasise the dual nature of television. Above all Williams emphasises the role of social agency and practice in determining the shape of the technology, as well as the particular forms of the culture. Thus Williams gives us a much more powerful set of conceptual tools with which to think about television than any of the wry epigrams of McLuhan (McLuhan 1964).

The broadcasting economy

Television and radio have been enormously important to the Australian economy. In 1954, shortly before his company applied for a commercial television licence, Frank Packer, owner of Consolidated Press, wrote to a business associate in Britain with a disarming question on what it cost to run a television station. In 1987 Packer's son, Kerry, sold TCN Channel 9 Sydney and GTV Channel 9 Melbourne to the Bond Corporation for $1 billion. Commercial television stations have until recently been enormously profitable affairs both in their own right and also in relation to the marketing and sale of a very wide range of goods and services in other sectors of the economy. Packer, Murdoch, Fairfax and others associated with the ownership of television stations in Australia have all become millionaires, and at least two of these figures, Packer and Murdoch, have become enormously powerful figures in the global media scene (McQueen 1977).

Most general accounts of Australian television (McQueen 1977; MacCallum 1967) focus systematically on the owners, the media moguls, implicitly asserting that ownership matters. It does indeed matter, not because this or that owner is interventionist in terms of dictating what kinds of content go to air but because, as I have demonstrated elsewhere (Tulloch & Moran 1986), there is a 'trickle down' effect whereby in an indirect but nevertheless determinant way, ownership affects the programming environment of television. The perceived environment becomes internalised as production norms, and production personnel find there are limits to what they can and cannot say on television.

For example, in the case of a drama series in 1985, I found that in the production of one particular episode of 'A country practice', 'Unemployment—a health hazard', the writer suffered political censorship when he suggested in the script that unemployment was structured rather than individual (Tulloch & Moran 1986). Although the producer who performed the censorship was partly motivated by more local concerns including the allegedly short attention span of the audience, the net effect of the filtration was to limit the kinds of explanation of unemployment made available in that episode of the serial. Thus the technology was at its most political even when it seemed to be at its most impartial, objective and outside the realm of politics in apparently innocently doing what it does most: entertaining the home audience.

Historically, the business of broadcasting, like other technologies of entertainment, has been monopolistic and oligopolistic. This has been one way in which business interests have safeguarded the massive amounts of capital tied up in this enterprise and helped to guarantee themselves a handsome profit. In this, as so often, capitalist interests have been able to use the machinery of the state at a series of levels in a variety of forms. Crucial here, in the case of broadcasting, was the dubious legal extension of copyright in 1915 in the United States Supreme Court. This saw the concept as applicable not only to literary but also to film, and later broadcasting, texts (Porter 1981).

Added to this in the 1920s was government intervention in the field of wireless to limit and control operators, thus putting thousands of small radio broadcasters off the airwaves. Governments in the United States, Britain, Australia and elsewhere even managed to produce a technical, engineering rationale (the scarcity of transmitting frequencies) as the principal motive for their action. By locating the reason in the realm of technology, governments mostly managed to escape charges of partiality.

Yet the technical reason was always spurious. This is suggested—in Australia and overseas—by the steady growth in numbers, especially in the 1960s, of first radio and then television stations. Again, too, in the United States the matter was doubly dubious. *The Radio Communication Act* of 1927 and the *Communications Act* of 1934 were a direct violation of the American constitution, specifically of the Fifth Amendment (Boddy 1979). The general moral is—as though it needed reiteration—that the technologies of home entertainment have always been more than simply machines invented by altruistic inventors motivated only by curiosity. The technology is innovated in a pre-existing field determined by pre-existing economic, political and social forces so that the technology comes to take a determinant form and shape because of their presence (Boddy 1979; Curran & Seaton 1989).

Cultural consequences

However, television and radio are more than hardware. If the state historically has intervened in the domain of broadcasting on behalf of business interests, it soon found that there were various kinds of interest which had different stakes in the technology. In particular, in broadcasting, television stations and networks are in the business of distribution and exhibition of television programming, while producers are in the business of making the programs.

The commercial networks have been Australian only to the extent that they have been owned by Australians, located in Australia and (generally) staffed by Australians. They have sought to maximise their profits by buying the cheapest material available. Coincidentally the Australian television stations came on the air in the mid-1950s at an historic point when much of American television production, particularly drama, shifted from east coast centres such as New York and Chicago to the west coast, Hollywood in Los Angeles; and from live transmission to being recorded on film. Until then television programs were not in a form which could be exported. With the programs clearing their costs in the American market, sales prices in Australia could be set at a fraction of the price that local producers had to charge for their programs.

The result historically has been a situation where both the commercial stations and the national broadcaster have imported large amounts of material. It took ten years or more for a viable local production industry to come into being and it occurred only because the imported programs provided an indirect subsidy for locally produced programs. Having listened first to the media moguls in setting the overall ground rules for the television system, the government from the mid-1960s on (and always with much prodding) developed minimum quotas for Australian content, particularly in the most expensive program form—drama. The argument was put and accepted from 1968 onwards that an Australian presence in film and television, brought about by direct government support and encouragement of the production industry, was a desirable policy initiative not only for economic reasons but also for cultural reasons.

In 1969 the federal government set about a scheme for the subsidisation of film and television production and, although the bureaucratic means as well as the levels of subsidisation have changed over the years, the support remains in place. However neither the economic nor the cultural reasons adduced for support of the film and television production sector finally make very much sense. In 1982 Sam Rohdie (Rohdie 1985) argued convincingly that Australian films and television have returned little to government bodies such as the Australian Film Commission on their investments. Instead, the support has become a permanent subsidisation of production out of taxpayers' revenue thereby lowering the cost of Australian film and television as far as distribution and exhibition are concerned. In the case of television programs, especially drama, government subsidy has meant that after 1969 Australian television networks had to pay proportionately less than they had earlier.

Nor have the cultural gains balanced the economic losses. Indeed there is the striking contradiction around the issue of national identity (Moran 1993). The cultural project of Australian film and television has been about the creation and circulation of a national identity and, even if the narratives have concerned rather more than stories of convicts and bushrangers, most Australian productions do subscribe to a monoculturalist Australia (Moran 1993). It is striking that one component of Australian government should be implicitly pursuing the dream of a unified Australian consciousness even while another sector should be propagating the ideology of a multicultural Australia.

However, since around 1987, Australian television has changed. Structurally the commercial networks have been weakened not only by a reckless inflation of prices for the buying of stations but also by the appearance of new rivals for their audience. Video has been a powerful competitor since 1980 and the imminent introduction of pay television will further weaken their economic base.

At the same time the production sector has been increasingly drawn into international arrangements in terms of such elements as finance, location shooting, and distribution. Australian film and television producers have systematically developed ties with international—especially American—distributors and this has led to a general mobilisation off shore. Thus, for example, the former Australian production company Kennedy-Miller is now based in Hollywood, while the Grundy Organisation's Australian operation is now merely one of 35 companies the group has worldwide with headquarters in the Bahamas. There has been a commensurate shift in the themes and foci of Australian screen narratives so that the founding impulse of a national identity released in Australian film and television is as impossible as ever.

To summarise, then, one might say that culture and economics are twin faces of the same coin. The cultural impact of television obviously matters enormously and I discuss some of the social, domestic as well as national consequences from other angles in some of the sections below.

Political influence

If radio and television have had a decisive impact on social and domestic life, their influence on twentieth-century politics has been equally profound.

By politics, I mean not only those machines of state that include the constitution, legislature, political parties and so on, but also those very processes that confer as well as deny political legitimacy on particular social activities. In this sense, the presence of the electronic media has been a powerful element in determining the social practice of politics in this century—and making it so very different from how it was undertaken in the nineteenth century.

The central political work of radio and television, at least in democratic countries, has been to sell the political system as a whole—including the presence of political parties, elections, the separation of powers, legitimate access to the media and so on—to the nation at large. While the electronic media have on balance not favoured the conservative political parties over the progressive parties, what has been favoured is the very notion of parliamentary politics as the only legitimate form of politics. In this sense, as various commentators note, there are clear homologies between discursive practices of this form of politics and the discursive practices of radio and television, particularly in the realm of news, current affairs, documentary and so on (Hall 1979). Beyond the role of the electronic media in eliding and dissolving contradictions of class, gender and ethnicity through their continued representation of parliamentary politics as equitable, democratic, responsive and effective, the media have also been an important instrument in delegitimating other forms of politics such as union actions (Glasgow University Media Group 1976) and the anti-Vietnam movement (Gitlin 1981). It is no wonder that, historically, western parliaments and political parties were happy to allow business interests to dominate the new field of broadcasting (Boddy 1979; Roberts 1969; and Briggs 1960), or that business interests were so keen to respond to these opportunities.

Using the model of the BBC in Britain, Australia committed itself early to the establishment of a public service broadcasting body—the ABC. Thanks at the time to a listener licensing system that involved paying a fee, these institutions cost little initially. However, from around 1945, the state has had to bear an increasing amount of the cost of public broadcasting. That it has done so with only minor murmurings points to the fact that radio and television are crucial technologies in the twentieth-century definition of acceptable politics. Even in a wholly commercial media system like the United States, radio and television continue to play the same central role of legitimation of the political system.

To understand the place of radio and television in countries such as Australia it is important to realise that broadcasting emerged in the 1920s and early 1930s, a time of economic, political and social crisis, not only here but elsewhere. The Australian economy had at the time been in recession for the best part of 30 years and in 1930 it was to embrace the Great Depression. The period of economic turmoil had seen the rise of rapid organisation and mobilisation of the working class in the new union movement and this, as well as more totalitarian political groups, posed a crucial threat to the existing political order. Broadcasting offered a new nation such as Australia, as well as older states such as Britain and the United States, the opportunity to achieve a new, hitherto impossible, national communications system which was crucial in the continued relegitimation of the state (Moran 1991).

The social impact

Radio and television have at different points in their histories been variously imagined as harmless and harmful in their effects. The 1927 Australian Royal Commission on Wireless dismissed the question of the social impact of broadcasting as hardly worth bothering with. However, by the time of the Joint Parliamentary Committee on Wireless Broadcasting in 1942 (Gibson 1942), radio broadcasting was envisaged as possibly having all kinds of influences, both good and bad.

The shift in perception about the power of the electronic media had come about not because there was hard evidence as to the powerful effects of broadcasting but because the radio industry and other groups, most notably politicians, had spent much of the previous decade claiming that broadcasting was a powerful tool in propagating messages. On the one hand, the radio industry emphasised to would-be sponsors that radio was a far more vivid, evocative and memorable means of advertising than were newspapers. Such a claim could not, and can never be, independently verified. However, the large sale of various consumer products such as Wrigley's Chewing Gum and Colgate Palmolive soap, heavily promoted on radio, seemed to guarantee its truth (Potts 1989).

Complementing this sale of consumer items was the radio selling of politics and politicians. In Germany in the 1930s it took the form of the broadcast of Hitler's speeches, while in the United States political radio seemingly took on an equally effective form in the shape of Roosevelt's fireside chats with the American public. Australian political thinking soon followed this pattern with the move in 1947 to the regular broadcasting by the Australian Broadcasting Commission of federal parliamentary proceedings. In such a move the technology of radio broadcasting was to be the instrument through which federal parliamentary politics was sold to the Australian public.

This phase of thinking about radio and later television lasted into the early 1970s. More recently though it has waned considerably. For example, a recent study by the Research Division of the Australian Broadcasting Tribunal of the child audience of 'A country practice' (Agardy, Burke & Wilson 1984) notes general patterns of liking, disliking and so on, and attempts to correlate some of these reactions with some demographic features of the audience. Implicit in the research and the strategies devised to accomplish it, including administered questionnaires, is the notion that the audience is free to do as it likes with a television program. The notion of powerful media and technology has been replaced by the notion of a powerful audience (Fiske 1989).

Yet I would argue that the research paradigm in this tradition of effects studies has been deeply flawed with sublime mythopoeticisation: the commercial broadcasting industries have simultaneously and contradictorily stressed both the effectiveness of the electronic media as an advertising medium and the sovereign capacity of listeners and viewers to switch off or otherwise ignore what is being offered. Instead, a far more useful way of thinking about radio, television and other electronic media such as video games, recorded music, and so on, is in terms both of the social meanings generated by the content of these media and the way these technologies of entertainment fit into the social fabric of everyday life (Silverstone & Hirsch 1992).

Television and radio as domestic technologies are often compared to a range of other white goods such as stoves, refrigerators, vacuum cleaners and so on. While the analogy is useful at one level, it ignores the social meanings or symbolism generated by the former technologies, an element entirely absent in the case of the latter home technologies. Quite simply, the content or programs of television and radio carry a set of social meanings or symbols that become part of the mental imagery of the audience.

This is not to subscribe to the notion that the audience passively receives whatever meanings are generated on the airwaves or the small screen. In television drama (Tulloch & Moran 1986)—an analysis that has been parallelled and extended by others in other areas such as the television transmission of rugby league (Nightingale 1992)—the television audience is actively involved in the construction of meanings. Yet this active construction occurs within determinant limits based on elements such as class, gender and ethnicity, as well as the range of positions and symbols offered by television or radio programs. Nor are particular audience members' readings of programs simply a reflection of the pre-existing social elements. Rather, audience members are actively involved in constructing themselves as, say, gendered subjects in and through their reading of television programs, just as they are in a series of other situations of work and leisure in both the public and the private spheres (Hobson 1981).

The technologies of home entertainment also impact on social life in a material sense. The radio receiver and the television set have taken a variety of forms, sizes and colours since the first broadcasting receivers were brought into the home in the 1920s. The changing physical nature of the object implied their intended location within different physical networks in the home at different points in time.

Early crystal set radio receivers were electrical gadgets of interest to electrical enthusiasts and likely to be located in a garage or attic. Some of the earliest of the complete radio sets in the late 1920s were grand furniture items in fine wooden mountings and intended for locations (as sources of home concert and ballroom music) in the lounge rooms of the affluent middle and upper classes. In the early 1930s much cheaper and more modest looking mantle radios appeared, and these were incorporated into more mundane areas of the house such as the kitchen and the bedroom. In the 1950s, with the replacement of the radio by the transistor, radios became miniaturised and portable. Listening became a more privatised, individualised and more mobilised affair. The commercial radio industry, eager to differentiate radio from a newly competing home entertainment technology, coined the slogan: 'Wherever you go, there's radio!'.

Beyond even the shape, size and imagery of radio and television, there is their physical imbrication into family life. Radio and television have systematically embedded themselves in both the physical space of the home and the rhythms of family life, and in the power and gender relations within the family. This imbrication of the entertainment technologies into the physical space, and power relations of family life has been a complex, on-going matter and can be usefully thought about in terms of the different program genres of broadcasting which presuppose different patterns of audience engagement, different audience members and roles, and different domestic activities in which different genres of programs are inserted.

Take, for example, working-class British housewives. They form a good percentage of the audience for 'Coronation Street' and 'Brookside', yet housewives are usually engaged in intensive domestic activities at viewing times. There are various compensating factors to allow engagement with programs while domestic activity occurs, for example old black and white sets in the kitchen, eating of meals, watching television, ironing in the living room (Hobson 1980).

Pejoratively dubbed soap opera, the form was developed by the American radio networks as a means of selling one particular product, soap and soap-related household items, to an audience that made crucial family-buying decisions, the housewife. The radio industry was anxious to fill daytime hours of broadcasting with sponsored programs, while advertisers were anxious to sell soap. To fit into the housewife's activities of housework, program producers developed narratives of romantic fictions which were low in action and high on conversation. To facilitate household activities, episodes were kept short while deliberate internal repetitions of conversation, plot incidents, characterisation, music and so on facilitated the pattern of listening while working. Most importantly the narratives continued across episodes so that listeners tuned in again and again.

The daytime television soap opera, the successor to the radio soap opera, continues on air down to the present. Indeed one program, 'Shining light', which began on radio in 1938 and which became a television serial in 1954, still on air, is the largest single text in the history of mankind. And despite the continuation of the daytime soap opera, the drama serial has also become a staple of prime time television programming in Britain, the United States, Australia, Italy, Brazil, India and elsewhere (Allen 1984).

Conclusion

Since the mid-century, television has become the central technology of ECI. At the same time, television has acted as a Trojan horse through which other modernising forces have affected nations on every part of the globe. It has done a great deal in shaping a common cultural and social outlook globally and has become a crucial element in the stimulation of consumption and distribution of manufactured goods and services.

Television has in the past been an important means of the reproducing and affirmation of a national cultural identity. However, broadcast television is in the process of a major transformation which results from—and gives rise to—a multiplication of home technologies of entertainment. Pay TV, cable, computer games and other innovative ECI technologies are in the process of forming new transformative links and relationships with the existing technology. The new technological combinations permit an increasing diversity of use, serving particular social needs and groups whereas, in the past and present, broadcast television has been capable of serving only a mass audience. Broadcasting television will be increasingly marginalised but more individual and small group needs are likely to be catered for. The present sense of national identity in countries such as Australia is likely to be eroded by a more diverse set of identities based on such elements as region, location, ethnicity and gender.

This chapter, then, has argued that the social, economic, political and cultural circumstances brought about by television are those of being plugged-in. Television—as the paradigmatic technology, typifying and also climaxing a range of apparatuses that have redefined in the twentieth century the relations between the public and the private sphere—has extended ever deepening roots into the very fabric of social life so that to think of television as only a technology is to miss the profound changes it has both imaged and triggered.

Where other elements of modernity such as the internationalisation of markets and of labour, and the development of ever faster and larger instruments of transportation, have stretched the political, social and economic fabric further and further, television has reinserted individuals into a social network. They are plugged into a constant stream of entertainment, communication and information. Above all, television carries the image of reality at a time when the legitimacy of older institutions, such as religion and politics, is being eroded. However, as I have suggested, the institution is in the midst of a fundamental change. Broadcast television is in decline even while video is becoming more ubiquitous.

References

Agardy, Susan, Burke, Jannina and Wilson, Helen 1984, *A Country Practice and the child audience*, Australian Broadcasting Tribunal, Sydney

Allen, Robert C. 1984, *Listening to soap opera*, University of North Carolina Press, Chapel Hill

Anderson, Benedict 1992, 'The new world disorder', *24 hours*, January Supplement

Barnouw, Eric 1978, *Tube of plenty*, Oxford University Press, New York

Barr, Trevor 1985, *The electronic estate*, Penguin, Ringwood

Boddy, William 1979, 'The rhetoric and economic roots of early American broadcasting', *Cine tracts*, vol. 2, no. 2

Bordwell, David, Thompson, Kristin and Steiger, Janet 1985, *The classical Hollywood cinema: film practice and mode of production to 1960*, Columbia University Press, New York

Briggs, Asa 1960, *Mass entertainment*, Fisher memorial lecture, University of Adelaide

Curran, James and Seaton, Jean 1989, *Power without responsibility: the press and broadcasting in Britain*, Routledge, London

Czitrom, Daniel 1982, *Media and the American mind—from Morse to McLuhan*, University of North Carolina Press, Chapel Hill

Dietrich Fischer, Hans 1979, *Entertainment: a cross cultural perspective*, Hanstrings House, New York

Dyer, Richard 1978, *Light entertainment*, British Film Institute, London

Fiske, John 1989, *Television culture*, Routledge, London

Gibson, W. 1942, *The minutes and report of the Joint Committee on Wireless Broadcasting* (the Gibson Report), Australian Government Printer, Canberra

Gitlin, Todd 1981, *The whole world is watching*, University of California Press, Los Angeles

Glasgow University Media Group 1976, *Bad news*, Routledge, London

Hadley, Patricia, McNulty, Jean and Salter, Liona 1979, *The mass media in Canada*, Lorrimer, Toronto

Hall, Stuart 1979, 'The unity of current affairs television', *Cultural studies*, no. 9

Hobson, Dorothy 1980, 'Housewives and the media', *Culture, media, language and society*, eds S. Hall et al., Hutchinson, London

——1981, *Crossroads: the drama of a TV serial*, Methuen, London

Kitross, John and Sterling, Christopher 1980, *Stay tuned: a history of American broadcasting*, Oxford University Press, New York

MacCallum, Mungo 1967, *Ten years of Australian television*, Sun Books, Melbourne

McLuhan, Marshall 1964, *Understanding media*, McGraw-Hill, New York

——1967, *The mechanical bride: folklore of industrial man*, Routledge and Kegan Paul, London

McQueen, Humphrey 1977, *Australia's media monopoly*, Vista Press, Melbourne

Moran, Albert 1992, 'ABC radio networking and programming, 1932 to 1963', *Stay tuned: an Australian broadcasting reader*, ed. Albert Moran, Allen & Unwin, Sydney

——1993, *Moran's guide to TV drama: a comprehensive directory of Australian series, 1956 to the present*, Australian Film, Television and Radio School, Sydney

Morley, David and Silverstone, Roger 1991, 'Communication and context: ethnographic perspectives on the media audience, *Handbook of qualitative methodologies for mass media research*, eds Klaus Jensen & Nicholas A. Jenkowski, Routledge, London

Nightingale, Virginia 1992, 'Contesting domestic territory: watching Rugby League on television', *Stay tuned: an Australian broadcasting reader*, ed. Albert Moran, Allen & Unwin, Sydney

Porter, Vincent 1981, 'The three stages of film and television', *Journal of the university film and video association*, vol. 1, no. 1

Potts, J. 1989, *Radio in Australia*, University of NSW Press, Sydney

Roberts, G.A. 1969, Early business interests and the ABC (BA Hons Thesis), University of New South Wales

Rohdie, Sam 1985, '*Gallipoli*, Peter Weir and an Australian Art Cinema', *An Australian film reader*, eds Albert Moran & Tom O'Regan, Currency Press, Sydney

Silverstone, Roger and Hirsch, Eric 1992, *Consuming Technologies: media and information in domestic spaces*, Routledge, London

Tulloch, John and Moran, Albert 1986, *A country practice: quality soap*, Currency Press, Sydney

Watson, Mary Anne 1990, *The expanding vista: American television in the Kennedy years*, Oxford University Press, New York

Williams, R. 1974, *Television: technology and cultural form*, Fontana, London

Annotated bibliography

Fiske, John 1989, *Television culture*, Routledge, London
A provocatively celebratory study of television. The author sees television as liberating as far as its audience is concerned, with the audience free to pick and choose the meanings it forms from television.

Glasgow University Media Group 1976, *Bad news*, Routledge, London
A ground breaking analysis of the demonic representation of trade union politics in the British media. These representations are always 'bad news', and the unions are invariably constructed as opposed to the public interest.

Moran, Albert 1993, *Moran's guide to TV drama: a comprehensive directory of Australian series, 1956 to the present*, Australian Film, Television and Radio School, Sydney
Useful guide to all series made in Australia and valuable overviews of the structure, development and programming of Australian television.

Silverstone, Roger and Hirsch, Eric 1992, *Consuming technologies: media and information in domestic spaces*, Routledge, London
A solid collection of essays that examine the domestic and social impact of technologies of entertainment, communication and information such as radio, telephones and computers.

Tulloch, John and Moran, Albert 1986, *A country practice: quality soap*, Currency Press, Sydney
A book-length study of the conception, development, production, circulation and reception of an Australian television drama series that attempts, at the same time, to demonstrate how the program is generally representative of television.

Williams, R. 1974, *Television: technology and cultural form*, Fontana, London
The best single book available on television.

ANTICIPATING TOMORROW:TECHNOLOGY AND THE FUTURE
Susan Oliver

As Australia—like most of the developed world—considers its economic malaise, politicians, economists and business people seek the answer to how can Australia dig itself out of the recessionary hole? But first we need to look at what sort of hole we have fallen into:

- Are we in a hole or are we really at a realistic plateau which reflects a low level of aspiration, a lack of will, an immaturity or incompetence inherent in our society and its structures?
- Is the hole simply a cyclical trough, a hiccup in our economic expansion, which will disappear as a matter of course in the not too distant future?
- Or, do we expect that by applying the policies of the past, we will somehow repeat their success and climb out of the hole using a tried and true path?

It is this latter definition of the hole to which we seem most attracted as we seek to emulate the policies applied in the growth period flowing out of the Second World War. Not only does this demonstrate a distinct lack of imagination but, I believe, we are also likely to find that we become stuck in the past while the forward-looking countries of the world will have moved further ahead as they grapple with those emerging issues which offer the most exciting opportunities over the next few decades.

So what approach would a competent nation take in order to answer the question of how to achieve a better situation than that which currently applies?

The answer has two components, both of which reflect the thinking and theory of futures studies and strategies. Directed at today's issues from a forward-looking perspective, futures work offers a powerful decision-making tool; one far more relevant than harking back to a past which we are most unlikely to replicate.

Component one: the inevitability of the future

The first component assumes a certain inevitability in the forces that have shaped society to date and which will shape it in future. This is similar in theory to the work of those economists who map the past through sinusoidal graphs which chart the short-term and longer-term waves of economics and market forces. They then extrapolate these graphs into the future. According to this theory, we are the victims of the impacts of these waves. They are the overriding force which determines the economic state of a nation.

These economists, however, fail to factor in one key variable, making much of their work fundamentally flawed. Their models deny the very existence of the phenomenon called change which is intrinsically part of today's expectations of the future. Yet the ever-increasing pace, and the complexity, of change means we are seeing fundamental shifts in the way we behave. Tomorrow's society will be very different, requiring very different responses to those which worked in the past.

As one of the most thoughtful authors in science fiction today, J.G. Ballard, says in an interview with *21.C* (Slaughter 1992, p. 95):

> the next 20, 30, 40, 50 years will be completely unlike those which rule our lives today and have ruled our lives in the past. We may move into a very dangerous and chaotic era, where all the old certainties and social cement that held society together will have gone. I think that there will be sudden quantum leaps in social values that would seem totally disconcerting to us if they happened now, but which probably our descendants in the next generation or two will take completely in their stride—just as we take completely in our stride sudden quantum jumps in social values that would have appalled or shocked our grandparents.

There is abundant evidence that change is already upon us. Although largely fuelled by technology and scientific advances, change is often driven by attitudinal shifts, the seduction of economic gain or by the search for higher living standards.

It is difficult to be fully aware of just how much change has already occurred, particularly over recent years. It is impossible to scratch the surface of most institutions now without finding enormous change, even upheaval, underway. The fate of the State Bank of Victoria, and the reasons behind that loss, strike at the heart of the institutions and value systems of Australia in the post-Second World War period. Respect for bankers, teachers, doctors and judges was an unassailable social given—once the demigods of society, the mighty are falling.

Thousands of government employment positions no longer exist, and Australian industry has experienced a major shake-out, not only due to the recession, but also as a result of labour market restructuring and deliberate efforts within industry, government and unions to achieve productivity improvements and international competitiveness: a far cry from the protected industry base of just a decade ago.

A shock wave in Australia resulted from the level of unemployment breaking first through the 10 per cent barrier, and then 11 per cent. Both political parties predict that unemployment rates could hover around 10 per cent well into the next century. And this is in a country which had full employment as its objective until the late 1970s, and at the turn of the century had the second highest living standard in the world!

At times a significant change or breakthrough is brought to our attention amidst excitement in the press and the community and we can recognise the form it takes—or will take—in our everyday lives. Such breakthroughs may be scientific, for example, in-vitro fertilisation or, some generations ago, the invention of television. Change is often insidious, however, becoming part of the expectations of the convenience and quality of our lives, without any heed given as to why or how we got to that point. Few people foresaw the dramatic turn-round in birth rates in Australia, as in the rest of the developed world, from 3.55 children per household in the 1960s to 1.85 by 1989. As the Australian National Population Council discussion paper in April 1991 (p. 20) explains it, this is a complex phenomenon which is:

> associated with fundamental changes in society including greater emphasis on individual well being and achievement and a general acceptance of a variety of lifestyles and behaviour. One aspect of this trend is the changing role of women in society and is reflected in the increased participation of women in the workforce. Demographically this is reflected in increases in de facto marriages, increased divorce and separation, later age at marriage, delay of the first birth, fewer children and voluntary childlessness.

Subtly, but markedly, a nation's set of values shifted over twenty years.

Technology forecasters take a rigid, simplistic view similar to the economic forecasters. They would have us believe in technological determinism; that technology advances inevitably, an independent abstraction removed from the world of human decision-making and social judgments. Technology forecasters, however, have to face some famous lemons, where technology has withered rather than waxed triumphant. The failed predictions for telecommuting, the paperless office, videoconferencing, the workerless factory and the cashless society suggest that these predictions were either techno-hype, techno-optimism or plain technocratic wishful thinking.

Clearly there is more than one force at work in determining whether or not society takes up the next scientific or technological breakthrough. Tom Forester (1992) called his paper 'Infollusion' and joined the many sceptics who believe that science enjoys a privileged world, secluded and sheltered from the realities of economics, politics and the needs of the person in the street. So many failed predictions, so many scientific discoveries with questionable advantages for humanity, heading away from a sustainable future.

But were these failed predictions techno-optimism? I rather believe in 'technology creep'. Society will achieve confidence and familiarity and, in due course, recognise the need for and greater convenience of many of the technologies currently on offer. This will be similar to the growth in usage of the fax machine and, before that, the word processor and the photocopier—each of which had to lean heavily on institutionalised practices and systems before they could find their way 'in' to everyday life. Do we imagine that we could do without any of these things now?

There is no doubt, however, that science and technology have advanced further than the social imagination, creativity and organisation required to capture their benefits. Yet what determines which science and technology is resourced? Can we assume that such research is driven by the current and emerging needs of society and industry, or is it driven by the skills and interests of the researchers? Unfortunately, the latter is too often true thereby further accentuating the gap between human needs and technological deliverables, and between scientific and technological know-how and human understanding.

Wendy Harmer (1991, p. 31) observes:

> Let's face it, most human beings reached their peak of understanding of technology in Ancient Egypt. In those days, most of us understood that if you hit things with a stick they broke, if you put a flame under them they burnt and if you pushed them they moved. A certain elite were acquainted with the principles of the pulley and the lever, but these boffins were considered the brains of the outfit. By the time the button, the dial and the switch came on the scene, the majority of the population had totally lost the plot. In fact as the 20th century draws to a close, 99.9 per cent of human beings cannot explain the scientific principle behind even the most humble of human inventions. Ask most folk to explain how a toaster works and they will be dumbfounded ... and I'm talking pre pop-up here.
>
> 'Um, well, the toast goes in the toaster and electricity comes in from a wire, which you plug into the wall and it cooks the toast.' Currents, filaments, wattage, amperage ... no idea. Silicon chips, programs, bytes, laser discs, memory ... you've lost me. Rocks, sticks, flames ... I may just about be able to help you on this one.
>
> Yet the demand for sticks and rocks has fallen considerably through the ages while the demand for the digital readout rages unabated.

Wendy goes on to imagine future archaeologists describing today's 'tribe' of people:

> How will you explain a tribe which sent the Hubble telescope to gaze on the plains of Mars yet needed instructions on a box of matches? A civilisation which could perform laser surgery on a foetus in the womb yet needed road signs which read: 'Form two lanes'? A society which split the atom yet put serving suggestions on cans of baked beans? (Harmer 1991)

If science has outreached the social imagination, creativity and organisation required to capture best its advantages, it becomes important to understand who the critical decision-makers are. We would appear to have a situation where the direction of basic or fundamental research is largely determined by the interests of researchers themselves; while the application of science and technology to the market place is largely determined by the distributor, utilising marketing and sales skills to convince an unwitting market that this new gadget is something no-one can do without. Witness the growth of the electronic games industry, higher resolution television and video recorders. The technology is marvellous, but the application of the technology makes little advance when the same old themes are rerun in merely a new technological guise.

Under these circumstances, it would appear inevitable that science and technology continue to plot their own course at an ever-widening tangent to the needs of the society they are supposed to serve. The general public will feel increasingly dissociated from the choices in direction which are available to it. And the march of science will reduce the chances of achieving a sustainable, equitable and just society.

If we understand this, it becomes possible to escape the clutches of fatalistic determinism; to move as a community towards a vision of the future, and play a part in realising it. This will, however, require *human* intervention, rather than reliance on the divine, free flow of market forces in an open economy. It will require the selection of a route other than 'business as usual'. It will mean that the community as a whole will have to become better educated about science and technology so, through understanding, they achieve greater control of it. It will demand that communities and governments go through the difficult, yet immensely exciting, process of choosing a preferred destination and charting a path to achieve it.

Choosing a future

In order to choose that destination or vision for the future, research is needed on the forces of change, unravelling the enormous potential for good that these forces can enable. In the end, real choices can only be made on the basis of the best possible information. A futures study would look at change across a range of issues, reflecting the complexity of interrelationships between them.

Social forces

'Social forces' here are used to refer to the value systems and the social and cultural behaviour of people. There is a wealth of information which indicates shifts in value systems and behaviour over past years, and points to future shifts in the next two to three decades.

Many argue that much significant change in the past resulted from a shift in social values. Everything—from the expansion of railway networks following the paths of the explorers, to the yearning to fly or the abolition of slavery—stands as an example of change, driven by social forces. Recent global value shifts include the extraordinary rise in concern for the environment and the trend towards increasing the status of women in society.

The uptake of technology is not driven by abstract notions of scientific progress or neutral concerns with efficiency. Take the invention of television, for instance. The inventor of television could see no marketable application for his product, which just goes to show how little scientists understand about the impact of what they are inventing. The uptake of television depended upon marketing, broadcasting technologies and program production; a complex commercial system enabled by the technology but beyond the imagination of the inventor.

Basically, it is social judgments and values which more often than not drive technological change. A beautiful example of this is the motor car, probably the invention which has had the single most significant effect on the shape of how we live in the twentieth century. In 1913, the fledgling Mercedes Benz company produced a forward projection looking at motor car usage in 50 years' time. They predicted that, globally, there would be two million cars on the roads, based on the rationale that it would be too difficult to educate large numbers of people to drive. By 1950 there were 50 million, and by 1989 555 million, vehicles worldwide.

Scientific and technological forces

The application of science and new technologies is the chief cause behind the exponential speed of change in the twentieth century.

Not surprisingly, the future development of technologies can never be based on precise prediction. In the important area of new electronic media, such as software, computers, home video systems, home entertainment systems and so on, the technological development is very rapid, and the industrial infrastructure in a constant state of flux.

As Miles, Cawson and Haddon (1992, pp. 67–81) discuss, product development is not proceeding along a single orderly path but rather along several paths, with indications that there is confusion among vendors as to the best path to take. This in itself is an obstacle to the greater take-up by the market place of new electronic media. The market place is playing a wait and see game as to how it all settles down. Buyers tend to hang back until the best technology is available at the lowest possible price with the greatest range of software training and back-up support. They want the technology that will sustain the longest product life cycle; and the technology with the greatest functionality and ease of use.

Industry structures and market place responses—combined with the complex interaction of product, of pricings and consumer expectations—are issues of considerable importance to the take-up of new technologies. Thus:

> Suppliers cannot impose their wills on markets, consumer feedback is most bluntly felt in the form of purchase decisions (and promoters of video conferencing and videotex will testify how market responses seem to defy logic), and opportunities for application of new technologies are as much a matter of the functions to which the technologies are applied and the system of habits, social attitudes and structures within which the function is currently performed.
>
> How far is lack of awareness of new technologies a significant factor hindering adoption—as opposed to costs, uncertainty over standards, faults with the products? We can agree that institutional factors affect the uptake and use of new electronic media, but it is important to recognise that these factors are not all on the user side of the equation. Suppliers may blame users for inertia, but often users will retort that suppliers have failed to produce the products that are needed—they are too over featured, too demanding of technical skills and back-up, not robust enough for the application and so on. (Miles & Haddon 1991, p. 12)

One of the greatest complaints made by observers of the development of new electronic media is that while the means of collection, transport and display of information has been improved, the actual material collected, transported and displayed is no better than it was. A relevant metaphor might be the construction of an expensive road between two villages, only to find upon arrival that the second village offers no additional benefits over the first. The road makers may be able to launch a vigorous defence of their well-engineered and reasonably costed roadway, but the user will perceive no advantages. It would seem logical that the very large financial resources needed to construct electronic 'roadways,' such as broadband ISDN, require that the range and quality of information and communication services which travel on the roadways offer benefits to users which warrant making the investment in the first place.

While futures planning tries to put today's technologies on the continuum of scientific logic, and study the very leading edge of today's research programs, clearly any technology forecast must include an analysis of interactions with the social, economic and political systems in which research directions are set. There needs to be analysis of how technologies are brought to the market and what factors influence consumers when they express their needs and preferences.

Environmental forces

Concern for the better management of environment is in its infancy in terms of altering the way industries, governments and individuals behave. A recent in-depth, nation-wide survey by Australian National Opinion Polls (1992) for the Federal Department of Arts, Sport, Environment and Territories found that the state of the environment was the number one, long-term concern for the great majority of people, taking precedence over other key issues such as unemployment and the state of the economy. Consumers are increasingly demanding environmentally-friendly products. Such is the growth in environmental awareness that it has become a mainstream issue affecting policy decisions made at the highest levels of government. Environmental management values, along the whole industrial chain, are gathering speed as countries change international standards and regulate to protect their environment. There is a world-wide move away from 'end-of-pipe' solutions towards cleaner production processes.

Saving the ozone layer by adapting industrial processes such as refrigeration and airconditioning, to handle chlorofluorocarbon (CFC) substitutes, is estimated to cost $200 billion alone. The global clean-up of toxic wastes has a similar price tag. However, the environment is both a cost and an opportunity. Environmental industries and technologies will become a major source of business, driving economic growth well into the next century.

Institutional arrangements

The way society has traditionally organised itself can impose barriers to change, and will influence the ability of people to respond to future change and opportunities. Solutions to problems can be predetermined by institutional arrangements as to who is asked to provide input. For example, when we look for the solution to more and better education, do we find the answer in more teachers, because that is what educational institutions are accustomed to providing; or more classrooms, because that is what our public works departments are in business to deliver? The answer may well lie in improved telecommunications, but the telecommunications interests would not be asked. If they were, their solution would be single interest and there would be a strong chance that the best solution might not be found. Why not? Because there would be no attempt to find an integrated solution by bringing together the creative forces of educationalists, technologists and students, and perhaps somebody from a completely different discipline. We can expect that solutions to problems in one industry will increasingly be derived from the logic of another industry altogether.

Future studies

While future forces can be categorised in this way, the most interesting and relevant aspect of futures work is the need to analyse the interactions and impacts of one category upon another—for example, the impacts of environmental issues on social values and behaviour; or the impacts of new organisations, like the European Community, on environmental management and technological development.

Utilising the hypothesis that one force will exert a dominant influence over another leads to the development of scenarios within which the individual, the organisation or nation can place itself and plan a response in answer to the question: 'How can we best position ourselves in that future time, in the face of inevitable change, to achieve our desires and needs? What policies should we make today, what actions should we take, to achieve that future position?'

Component two: a strategic vision

The second component of a competent approach to the future builds on the first. It asks questions such as 'What do we want the future to be like? How are the fundamental needs and value systems of individuals and communities reflected in the future we achieve?' This is usually termed a vision, or a strategic vision.

If an individual is asked to describe their personal vision for their own future, their response covers a range of issues usually to do with personal relationships, economics, career and professional satisfaction and recreational interests. It is of concern that these values are not reflected in the objectives of governments whose goals are much narrower, and usually weighted in the area of economics and its various indicators. The need to be re-elected dominates all government policy-making providing a short-term, rather than a strategic view.

A strategic vision for a nation needs to sit within the global context and address the broadest range of human values and needs. At the next level, a strategic vision for an organisation, or an industry sector, sits within the context of the national vision and derives benefits from the commonality of purpose. While this sounds extraordinarily organised, a model which approaches this is that of Japan. Certainly the Japanese vision reflects economic and industrial development objectives, in harmony with cultural values which accept the need for long-term planning, with the state serving and working with industry to achieve its objectives.

A vision for work

The abandonment in the late 1970s of the government's objective of full employment, for example, has had considerable ramifications for Australia's economic and cultural values. Economically, labour market restructuring and the recession are causing a much needed shake-out in Australian industry, resulting in greater competitiveness of surviving companies. While the inefficient companies have gone and higher levels of productivity are being achieved by the surviving companies, many jobs have disappeared, probably forever.

This upheaval has not been adequately addressed, however, at the cultural level. Work has a much wider importance than its economic value. Work is the key not only to how a person's status is defined, but it also provides the bridge from school to adulthood. Consequently the impacts of the loss of jobs during the current recession—reflecting a cost of an unintegrated strategic vision—are being felt disproportionately by fifteen-to nineteen-year-olds. This has led to higher retention rates in education, but also created the spectre of a group of people in society who may never find a job in the traditional sense.

The 1992 Prime Minister's Youth Summit discussed such issues as on-the-job training, and a youth wage, but the debate about skills levels achieved by young people (and their affordability) denies the real issue: our economy is currently unable to create the jobs required. Hence the Youth Summit also went on to define new types of jobs in community projects with linked social, environmental and economic objectives. We are beginning the process of redefining work—an essential task in redefining our strategic vision.

How work is defined will determine the sort of society we have in the future. Work experience, learned skills, and the sense of belonging which work engenders, provide the basis for the development of human beings as citizens. The more complex the work and the greater the level of involvement a person has with their work, the greater their ability to engage in making the difficult decisions which citizenship increasingly demands: 'at the deepest socio-cultural level, the degree of complexity of a person's work and their level of involvement in decision-making determines to what degree that person is capable of engaging in a society' (Wettenhall 1992, p. 76).

The need for society's goals to embody a range of human values

Ian Miles discusses the narrowing to economic concerns of the success (or otherwise) indicators of government policies and programs (Miles 1985, p. 10). He highlights how development has taken place as 'more or less rapid movement along a single dimension of progress'. This is consistent with national financial management policies where:

> the world economic crisis was a reason for discarding concern with questions of social welfare and environmental protection: it was time to reiterate the 'magic', and re-impose the discipline, of the market place, in which such problems would find their own solutions almost automatically. (Miles 1985, p. 2)

But solutions have not happened automatically and Miles terms the outcome 'mal development':

> It is clear that the so-called 'developed' nations are far from being in a state of timeless perfection, and that the alluring models of material progress they offer the rest of the world are seriously flawed. It is likewise evident that the term 'developing nations' is inappropriate to many of the countries to whom it was so readily applied in the recent past. In many respects, the Third World seems to be importing most of the social problems of the industrial countries, but failing to acquire much of their material affluence. Increasingly, attention has been drawn to the linkages between material prosperity in one world region and poverty in another; and those between the emphasis on satisfaction (and unlimited extension) of material needs and the discounting of more social relations, political and non-material aspects of well-being. Across and within countries the world displays an inequitable and lop-sided pattern of development and repression of human potentialities. It is maldeveloped. (Miles 1985, p. 10)

Miles is just one of a number of international futurists who argue for the wider spread of human values to be represented in setting our goals. His working group suggests a range of indicators of importance to the functioning of society within the context of the world as a whole. His research team gives five goals.

- *[achieving] a harmonious relationship between persons, society and nature:* insuring the fullest flowering of human potential without degrading, despoiling or destroying society or nature . . .
- *social equity:* [insuring] human development is equally possible for all members of the society;
- *inter-regional and international equity:* society permits and promotes human development of its members along with respect for the integrity of [other] societies (i.e., no economic exploitation, political domination and/or cultural oppression which prevents the members of other societies from achieving their human development);
- *living presence of the future:* human development of present generations is not to be pursued at the cost of endangering the existence and development

of future generations. We refer especially to preservation of the [natural] environment and also to respect for (not submission to) historical achievements and values which help to define people's cultural identity;

- *sensitiveness to the present:* nevertheless, the development of future generations must not mean the deprivation of the present generation. Construction of a human future is a condition for a human-centred development process, but oppression of people in the name of a distant future can not be justified. (Miles 1985, pp. 11–13, layout modified)

This world view tries to address the issue of uneven development between countries, the uneven access of nations to technology and the fact that many societies and groups are locked out of decision-making altogether. As the old saying goes: you can't control what you don't understand. Miles believes that technological dependence:

refers not only to the need of countries or regions to import technologies, but also to their inability to mobilise sufficient capabilities to apply these technologies to their development needs.

As the Junta del Acuerdo de Cartegena [1976] notes, the consequences of such dependency can include

(a) the loss of control of decision making in programming, production and marketing;

(b) the frequent import of inappropriate knowledge . . . and

(c) the reduced negotiating power of member countries in the purchase of technology. (Miles 1985, p. 190)

Richard Eckersley (1992) points out that it is clearly in the interests of educators to increase the level of science and technology taught in the classroom. In future, children will need to have a much better grasp of science and technology if they are to harness its utilisation for the tasks of creating a more sustainable and equitable society. To take a further example of another social group, women, universally under-represented in decision-making roles in science and technology in general, Miles writes that this is a simple matter of equity:

an important section of the population is marginalised in terms of one of the key development-related activities. It is likely to benefit less from the rewards . . . and to have less opportunity to make any of its own distinctive concerns felt in the overall direction of science and technology and correspondingly, to understand the options that are in principle available when it comes to the final use of the technology or its products . . . [M]any of the more deformed growths of modern science and technology—especially in the military sphere, but even at the level of the world-views of natural science—reflect the dominance of patriarchal culture in science and technology; which is facilitated by the numerical dominance of men in the science innovation system. (Miles 1985, pp. 191–2)

Other social groups have poorer access to the 'science innovation system' and some national groups have virtually no access at all. Those who followed the United Nations Conference on Environment and Development in Brazil in June, 1992, will be aware of the 'North–South' divide and the bid by third world and developing nations for access to environmental technologies and for free technology transfer from industrialised nations.

But what should be appreciated is that *only* if we freely transfer the new environmentally-friendly technologies can we hope to rein in drastic increases in the rates of pollution as foreshadowed by the Club of Rome in their seminal work *The limits of growth*, now updated (Meadows et al. 1992). At present the 20 per cent of the population living in the developed world uses 80 per cent of the resources consumed by humans. It takes little imagination to foresee the severe environmental and health problems if we do not move towards less wasteful, 'cleaner' industrial production processes—both in the north and south.

Conclusion

If social and cultural values are not expressed in the society we create, dysfunction will occur in the form of undesirable breakouts such as crime, violence, suicide, mental illness. Modern western society already bears witness to ample evidence of this dysfunction, as Eckersley (1992) highlights. On our current course, the future would seem to be trending inevitably towards larger police forces, greater expenditure in scientific R&D for crime detection and surveillance, bigger riot squads and new technologies for crowd control.

The principle of first building a strategic vision points us in the direction of understanding the underlying, fundamental values and needs of individuals and communities and building these into the objectives and accountability of government, and of business. Such an understanding is incorporated in the five social indicators suggested by Miles and his team (above). Just as a strategic plan focuses the allocation of resources within a company, a strategic vision focuses the attention over a much longer period and builds in the flexibility to shift tactics in the wake of inevitable change. In other words it offers flexibility within a charted direction. If the strategic vision for a company reflects the range of aspirations within the company and plausibly reflects the key competencies and commercial relationships of that company, it creates a sense of purpose and a unity in that purpose.

A strategic vision is not only a statement of what an organisation or a country will be in the face of inevitable change, but also—by taking part in directing change—a determinant of getting where it wants to go.

References

Australian National Population Council 1991, *Population issues and Australia's future*, Population Issues Committee, Australian Government Printing Service, Canberra

Eckersley, R. 1992, *Youth and the challenge to change: bringing youth, science and society together in the new millennium. Apocalypse? No!*, Essay No.1, Australian Commission for the Future, Melbourne

Forester, Tom 1992, 'Infollusion', *PC User Australia*, April

Harmer, Wendy 1991, 'Gadget, greed's ghosts', *21.C*, Winter/Spring, Australian Commission for the Future, Melbourne

Junta del Acuerdo de Cartegena 1976, *Andean Pact technology policies*, International Development Research Centre, Ottawa

Meadows, Donella, Meadows, Dennis and Randers, Jorgen 1992, *Beyond the limits, global collapse or a sustainable future*, Earthscan Publications, London

Miles, Ian 1985, *Social indicators for human development*, Francis Pinter, London

Miles, Ian and Haddon, Leslie 1991, *Towards a paperless society*, PREST (Program of Policy Research in Engineering, Science and Technology), University of Manchester, Manchester

Miles, Ian, Cawson, Alan and Haddon, Leslie 1992, 'The shape of things to consume', *Consuming technologies: media and information in domestic spaces*, eds R. Silverstone & E. Hirsch, Routledge, London

Slaughter, R. 1992, 'Psych-fi: J G Ballard's startling visions of the future', *21.C*, Autumn, Australian Commission for the Future, Melbourne

Wettenhall, G. 1992, 'The great divide: rethinking the future of work', *21.C*, Winter, Australian Commission for the Future, Melbourne

Annotated bibliography

Harms, L.S. and Sabado, F.S. 1991, *Teleworking and telelearning: perspectives from the Pacific*, Pacific Telecommunications Council, Hawaii

This collection of papers representing authors and perspectives throughout the Asia–Pacific area indicates ways in which 'technology creep' is integrating telecommunications with people's work and education; and points to future trends in this field.

Miles, Ian 1985, *Social Indicators for Human Development*, Francis Pinter, London

This book questions the use of Gross National Product as the indicator of 'development' where it refers to industrialisation and modernisation, but not necessarily to the true needs and expectations of people. Miles argues that economic growth in the way it has been achieved by Western industrialised societies ignores truly desirable goals across a much wider spectrum of considerations.

Meadows, Donella, Meadows, Dennis and Randers, Jorgen 1992, *Beyond the limits, global collapse or a sustainable future*, Earthscan Publications, London
This is an update of the Club of Rome's earlier report *The limits of growth* which was originally released in March 1972. The book argued 'zero population growth' and urged environmental management. *Beyond the Limits* was released in 1992, at the time of the United Nations Conference on Environment and Development in Brazil.

TILL DEATH US DO PART: TECHNOLOGY AND HEALTH

David More and Elizabeth More

Images of dramatic technological progress dominate our understanding of modern medicine . . . As a culture we are fascinated with the details of medicine's most recent miraculous advance. The limits of technology seem boundless. Although increasingly aware that progress sometimes occurs at significant cost, both social and economic, we await eagerly news of the latest test-tube baby or liver transplant . . . Clearly, the use of technology cannot be independent of its social context. Especially in the case of medical technology, with its potential for evoking strong feelings carrying potent symbolic references to the body, life and death, the relationship between the machine as object and its user is multifaceted (Koenig 1988, pp. 465–7).

In a technologically biased society, the assumption is generally made that a new technology will be an improvement, and the onus of proof is more strictly put on those who doubt its usefulness than on those who claim its benefits (Bates & Linder-Pelz 1987, p 119).

As in many other arenas of contemporary life, the use of technology in the health care industry is increasing. Emerging streams of medical technology include, for example, medical artificial intelligence, biosensors and implantables, diagnostic imaging, genetic engineering, home health/self-diagnostics, new-wave laser surgery, office automation, super drugs, transplants and implants (Coile 1990). In spite of our modern imagination often being awed by the power of medical technology, there is public concern that the implementation of new technology may be one of the main factors in escalating health costs. Moreover, the implications of technological developments for a variety of areas—for health professionals; for the political, economic and legal systems; and on social values and ideals—are on the research agenda.

This chapter puts forward two key ideas. The first is that there is a rational approach to the evaluation of health technology and the second is that forces shaping the development, adoption and overall use of health technology are considerably more complex than may first be appreciated. These include a combination of inappropriate incentives, patient expectations, the politics and economics of the industry, and inadequate comprehension of the issues. The lack of processes and mechanisms to ensure the adoption of a rational approach is essentially a social and political problem which reflects the complexities of human nature faced with the issues of life and death.

Health care delivery in industrialised societies

It is important to recognise the nature and scope of health care delivery in modern industrialised society. The scale of the undertaking can begin to be appreciated when a few key facts are acknowledged. Examples are:

*   The British national health system employs over one million people.
*   The United States is projected to spend approximately 14 per cent of its gross domestic product on health in the early 1990s (amounting to about $US800 billion). Informed estimates suggest that this figure could reach $US2 trillion by the year 2000 (Schwarz 1992) unless there is fundamental reform of the US health delivery system. The historical rate of rise of health care expenditure in the US has been dramatic, moving up from of the order of only 4 per cent of GNP in the 1960s. These projections are unfundable and the US health care system has emerged as a major domestic political issue. The concern is magnified by the realisation that, despite proportional GNP expenditure of almost twice that of other industrialised nations, the net health status of the US population is not measurably better and, indeed, on some macro indices is worse.
*   Australia spends in excess of $25 billion on health care amounting to of the order of 8 per cent of GNP per annum. Here, too, the increasing politicisation of health care is of concern.
*   The newly industrialising nations of Asia and the Pacific Rim are experiencing rapid growth in proportional expenditure on health care as their capacity to afford care rises and the expectations of their populations increase.

Activity on this scale reflects the importance of health care in most national priorities, but also highlights the social, political and economic importance of debate relating to the mechanisms for, and the efficiency of, health care delivery.

Of total health expenditure in industrialised societies between 65 and 75 per cent relates to staff and administrative costs, part of which is directly related to the sophistication of the technology employed. The remaining percentage is predominantly the cost of the various items, varying from the most basic to the most 'high-tech', used in the delivery of care. Some other points include:

- Access to quality health care is seen as an inalienable right in most industrialised societies and is always a politically sensitive issue.
- The scale and importance of health care delivery makes change very difficult, and the large number of stakeholders makes obtaining agreement to directions of reform extremely complex.
- Historically, the fundamental nature of health and the characteristics of the industry have meant that medical applications of any new technology are always one of the first to be developed.
- The rising cost of health care worldwide is provoking considerable social and political concern and virtually every conceivable mechanism to contain these costs is being considered, implemented and evaluated somewhere in the world.

It is a central contention of this chapter that appropriate development, adoption and use of advanced medical technology is not a major contributor to the escalating costs of health care being seen in all industrialised and industrialising countries. However, it is clearly recognised that the rational framework outlined in the following pages is not by any means followed as new technologies emerge and that this failure has resulted in substantial and unjustifiable use of what are now becoming scarce health finance resources.

The inappropriate application of medical technologies

The factors listed below are all significant contributors to the inappropriate application of advanced technologies:

- Patient expectation of receiving the latest and most 'high-tech' intervention conveniently near to home, leading to excessive duplication of expensive equipment in many more sites than can be justified, if it is to be fully utilised.
- Medical demand to have easy access to all the latest equipment so as to serve better the patient and demonstrate how up to date the doctor is. This is often coupled with the desire on the part of practitioners to remain current by making sure they are familiar with every latest technique or technology.
- Aggressive marketing of each new technology by the developer (often aided by mass media agendas), in an endeavour to ensure profits are made and development costs are recovered. This is especially relevant in the marketing of pharmaceutical products, where simple, cheap, effective medications are often replaced by marginally better products at dramatically higher prices.
- The failure of practitioners to stop the use of older outmoded techniques as new and more effective ones come into existence. An example of this is the continued use of barium-based radiology long after

endoscopic techniques have been shown to be both more reliable and less unpleasant for the patient.

• Pressure in many health systems to get a good return on investment from a piece of technology, leading to administrative pressure to make sure the investment is profitable, and that the technology is extensively used. This is a common issue in the US health care system. In Australia a similar incentive may exist to employ an expensive technology if there is a fee for the practitioner associated with the use of the technology.

• A degree of medical denial of the importance of consideration of issues such as cost when planning patient care. Medical education and clinical practice are yet to develop real ways of factoring in such issues into the clinical decision-making processes.

Scientific rational decision-making is but one component of the forces that drive the application of health technology. With the wrong incentives in place, profit and other such motives, or decisions driven by the emotional overtones affecting the industry, can have a significant effect on the overall cost of health care. Assessment is no simple matter.

For example, less than 30 years ago the world witnessed the first 'successful' heart transplant with the patient surviving for two years. Today, the survival rate in adult transplants is 80 per cent alive after five years. In Sydney, for example, about 60 such transplants are carried out annually while the Melbourne Children's Hospital has done 24 heart transplants since 1988. These procedures are virtually 'routine' after their initial fanfare in the 1960s. But such transplant organ procedures have also killed—spreading AIDS and other diseases through transplanted organs and, more recently, encompassing the development of an often abhorrent international trade in human organs, essentially with the first world supplied by the third (Ragg 1993).

Quality and cost of health care delivery

Most trends in health care delivery have as their objective the reduction of the total cost of treatment with the preservation of as much quality in the overall care as possible. It is important to recall that quality, in and of itself, can reduce cost. For example, surgery performed correctly the first time with modern technology can reduce the number of complications and the associated costs. Among key trends are:

• reduction in the emphasis on expensive institutional care and a maximisation of the care delivered in a community or outpatient setting;

• an increasing emphasis on the performance of medical and surgical procedures on a 'day only' admission basis rather than as part of a normal hospital admission;

- introduction of significant work-place reform and restructuring within health care facilities;
- an increased use of information technology to enable effective monitoring of costs, quality, outcomes and output;
- increasing emphasis on healthy lifestyles, health education, and preventative aspects of medicine;
- the use of techniques such as 'case mix management' to accurately identify the clinical products produced by a hospital and to ensure that these products are produced at a reasonable cost.

Against these reforms and initiatives there are a number of factors, some technological, that are driving the continuing rise in health expenditure worldwide. These include:

- the progressive ageing of the populations in most of the industrialised world with the increase in resources required to care for the patients;
- health costs associated with the increase in substance abuse and the AIDS epidemic;
- the continuing rise in the populations of most of the third world associated with improvements in health education resulting in increased demand for more and more sophisticated and costly services;
- worldwide escalating expectations of patients;
- the rising cost of labour worldwide, with the huge impact that has on total health expenditure, especially in administrative and professional costs;
- the continuing development of new technological techniques and equipment for use in the health care delivery process, with the inevitable associated cost increases, including that of specialist staff.

In summary we have spiralling need and expectation, increasing technological promise and rising costs, all of which should be balanced against the capacity and willingness of nations to fund the ultimate bill and the issue of balancing benefits to society overall. As one writer puts it (Schwarz 1992, p. 2):

> The only limit to what scientific medicine can accomplish seems to be our ability to pay for it. Organ replacement, gene therapy, and some of the new reproductive technologies are examples of techniques for which both the applications and the costs appear to be limitless.

Constructing a social understanding of a medical technology

There are many stakeholders in the process which results finally in the use of a new health technology. Among these are the technology developer, the manufacturer, the distributor, the user, the community and the patient.

Consideration of any particular technology will reveal widely varying inter-ests, incentives, motives, and different social constructions of reality among these stakeholders. Furthermore, the final outcome, in terms of eventual use, is often determined by the power inherent in the various stakeholder positions and the methods by which various interests are ultimately resolved.

Moreover, there are continuing social processes that contribute to the operation of a technological imperative in modern medical practice—the process by which technological interventions become the only acceptable course. This is clearly detailed in Koenig's (1988) analysis of Therapeutic Plasma Exchange (TPE). Koenig focuses on how individuals construct understandings of complex technologies affecting their lives and how the meaning of medical technology changes with the use of new machines. She points out the continuum between technologies from pure experiment to the standard of care, asserting that transformation of meaning is an inher-ently social process which serves to sustain the technological imperative. Koenig's study also demonstrates how:

- our underlying cultural preoccupation with mastery of the body through increasingly sophisticated technology is supported by important political and economic considerations;
- once technology is developed, the formidable forces favouring adoption and continued use as standard therapy often mean that we have widespread clinical practice diffused before clear evidence is available concerning the actual usefulness of that technology;
- evaluating a new technique as a standard therapy cannot be wholly achieved in terms of results but must derive from a complex reading of the social setting within which it is used; and
- the difficulty of preserving individual patient autonomy faced with the moral imperative to make use of new techniques.

As Koenig (1988 pp. 486–7) suggests:

> even in a highly rational, scientific setting the meaning of actions and events evolves, at least in part, from the underlying social and cultural organisation . . . Scientific studies of effectiveness, assessments of economic costs and benefits, and political considerations all have their place. Although often ignored by policy analysis, an understanding of the social context of technological innovation can inform our knowledge of the meaning of new technologies.

Her study highlights the following (1988 p. 490):

> Since high technology medicine seems to be a direct embodiment of scientific knowledge, we wish to believe that the application of these new machines to patients is objectively determined, comprehensible to all. This case study of TPE reveals that even in the seemingly rational world of medical science one

cannot ignore the social realm—encompassing the highly subjective experience of participants in medical innovation. A full understanding of the relentless advance of medical technology requires knowledge of the social world in which medical machinery is developed and used. As routinisation occurs and a new meaning for a medical technique solidifies, policy options narrow.

Such alternative evaluation aids understanding of the rationale for choices made among competing demands for limited resources, especially in terms of appropriately balancing high technology as against other services—for example, funding increasingly sophisticated artificial heart transplants versus research into coronary artery disease itself and preventative medicine. Given that, in most Western societies, 'medicalised' definitions of the meaning of health and illness, life and death, have become paramount (Arney & Bergen 1984), alternative evaluation is indeed a difficult task.

Do medical technologies lead to better health?

Health status on a macro scale is usually assessed using a series of indices which allow for comparisons across both nations and time. Typically the indices used include life expectancy, infant deaths per 100 000 live births and maternal deaths per 100 000 births. In addition to these general indicators, mortality rates to specific illnesses, usually within specific age ranges, can add further valuable information. Examples may include deaths from gastroenteritis in children less than one and deaths from tuberculosis in adult males. When such indices are examined, the impact of expenditure on medical technology is often difficult to detect.

The era of specialised technological medicine can reasonably be said to have begun with the invention of the X-ray in the closing years of the nineteenth century. Thereafter there have been, among other innovations, the development of antibiotics, body fluid analysers, CAT (computerised axial tomography) and MRI (magnetic resonance imaging) scanners, nuclear medicine, ultrasound imaging, biotechnologies and genetic engineering and the development of a legion of new pharmaceutical products to treat or control a wide range of disorders. A range of material sciences developed in parallel, have led to all sorts of replacement parts for the human body from intra-ocular lenses to artificial hip and knee joints.

In virtually all indices measured, however, trends toward better health started with improvements in literacy, sanitary engineering, housing and nutrition that began some hundred years before the widespread use of any specialised medical technologies. The rates of improvement in many cases have shown very little change despite introduction of demonstrably proven technological treatments. As Vimpani (1991) points out, a very large proportion of all the moves towards healthier populations seen in the twentieth century can be attributed to superior housing, education and nutrition, with only possibly 10 per cent of the change attributable to medical care and all the technologies associated with it. This must be considered in the light of estimates that suggest that as much as 50 per cent of the rise in the costs of health care delivery since the 1960s can be attributed to the application of technical innovation.

All the above does nothing to diminish the importance of an effective technological cure to the individual concerned, but it does raise the issue of the need to establish valid cost–benefit and efficacy indicators for a community for a new technology before substantial investment is made and it becomes incorporated in clinical practice. This becomes an imperative in circumstances such as those pertaining in the United States where the Secretary for Health and Human Services felt impelled to write that, 'The cost of health care now threatens to overwhelm and engulf our nation' (Sullivan 1991, p. 10).

In some situations, where new medical procedures, such as vaccines, have demonstrated overwhelming improvements on past practices, assessment is less of a problem. But often, evaluation of technological innovations in procedure and equipment is much more complex to undertake.

Medical technology and escalating health care costs

With the need to control the spiralling costs of health care delivery, pressing hard on all industrialised economies, it is only natural to see the rise of interest in methodologies designed to assess each new technological advance. Patients should be well served, and there should be a positive overall effect of that technology on the net health status of the community. The overall cost–benefit of the technology under review should also be factored into the assessment to allow rational debate on the final place the technology should have in the overall health delivery process. Such debate is often difficult as we have seen in recent controversy surrounding techniques and technologies involved in fertility control—for example with in-vitro fertilisation and the way in which such technology is now used to allow post-menopausal women to have children. Some might wonder whether this is the use of medical technology gone mad.

Those placing the blame for the escalating costs of health care on advances in, and uses of, recent medical technology do so for reasons such as the following (Jecker & Pearlman 1992):

- Many in the field believe technically curative medicine costs too much compared to other uses for the health care dollar, especially in the area of preventative medicine.
- Scientific evidence may be less effective than subjective factors such as

the prestige of new technologies and advancement of professional careers, in determining clinical practice and research.
- Research orientation, especially in academe, is focused on expertise in the latest technologies.

An emerging concern is the increasing dependence by medical practitioners on technical experts outside the medical community. While the concerns outlined above suggest the possibility of an unquestioning technological imperative in contemporary medicine, problems with a simplistic approach emerge, including the following:

- technology sophistication itself is not an adequate measure of what health care society is morally obliged to provide;
- technology per se is not necessarily unwarranted; rather its unwarranted use may be the problem;
- taking a historical view, we can see that what is high-tech today is low-tech tomorrow.

Moreover, well used, some technologies result in the reduction of health care costs. For example, home treatment instead of hospital treatment of young asthmatics with home nebulisers can save many thousands of dollars; and a more sophisticated air-powered staple gun means orthopaedic surgeons can cut significantly the operating time by stapling bones in place rather than using screws and the like, thus saving substantial health care dollars (Robertson 1991).

Also of interest are the costs and effectiveness of the technologies being replaced. In some circumstances when this comparison is made, an 'expensive' new technology saves money while improving the effectiveness of therapy. The introduction of Histamine Receptor Antagonists (e.g. cimetidine) for the treatment of duodenal ulcer is an example of this. When introduced these drugs were dramatically better than any other therapy for the disease. They converted an illness which required weeks of hospitalisation to a disease that was effectively able to be managed in an outpatient or office setting. The expense of the drug in this situation was offset not only by the hospitalisation costs avoided but by the improved symptomatic control of the disease, allowing sufferers to return to productive work in days rather than weeks following diagnosis.

A rational approach to the evaluation of health technology

A number of factors should be considered in assessing a new technology.

Efficacy

Does the technology work as intended?

Approaches to assess this issue are varied but probably the most useful and scientific is Randomised Controlled Clinical Trial (RCCT). Ideally such a trial uses the best elements of scientific experimental design and statistical analysis to evaluate a proposed new technology against the existing state-of-the-art approach in a sufficiently large number of patients to obtain a statistically valid and credible result. It will be appreciated that such effort is expensive, time consuming, and complex, and is usually undertaken when preliminary evaluation has determined a high chance of a positive result.

Increasingly it is evident that governments are requiring such studies to be performed before widespread use of any new technology is funded from national budgets. Indeed in many countries there are now well developed regulatory bodies to ensure the scientific validity of the information provided by the technology manufacturers (e.g. the US Food and Drug Administration).

Although the rigorous scientific approach is conceptually ideal, in some circumstances it may not be possible. A recent example of the recognition of this is the US Food and Drug Administration's introduction of a 'Modified Treatment Investigational New Drug' application process for new drugs which may have an effect on fatal diseases such as human immunodeficiency virus (HIV) infection and some forms of cancer (Hendee 1991).

While the RCCT is the 'gold standard' by which efficacy of a technology is assessed, it is not always appropriate (e.g. it is impossible to do a double blind trial of two competing imaging technologies when the same staff will be assessing obviously different images from the two different imaging modalities). In such situations other less robust techniques, such as performance analysis, case studies, historical trend review, etc. may have to suffice.

Risk–benefit ratio

For a technology to be seen of value, not only must it work, but it must be shown to have a degree of safety consistent with the threat posed by the illness it aims to investigate or treat.

For simple self-limiting ailments (such as a cold), it is clear that virtually no risk is acceptable. On the other hand, for illnesses that have a high risk of patient death, especially in the young, virtually any risk may become worthwhile.

Reasonable cost–benefit ratio

Taken as a whole, any new technology should be able to demonstrate a reasonable cost–benefit ratio.

While the cost side is usually easily assessed, there is more difficulty in assessing benefit in overall financial terms. The difficulty lies in the attribution of a financial value to such factors as quality of life, rapidity of cure, associated morbidity and, ultimately, the value of life itself. This is similar to the difficulties evidently experienced with cost–benefit analysis, sustainable development and the environment. In the final analysis the judgment of ultimate cost–benefit is a moral and ethical issue based in societal values, not a pure financial calculation.

Positive net effect on community health status

Technology exists to serve community needs. With this perspective it may be difficult to make purely scientific and rational assessments within the health domain. Community attitudes and perceptions, as well as factors outside the purely clinical domain, can result in adoption or deployment of technologies, which is not strictly justifiable in clinical terms. An example of this may be seen in the proliferation of CAT and MRI scanners in the United States in numbers well beyond those required to meet the clinical needs of the population.

Clearly, medical technology assessment involves a comprehensive form of evaluation in which short and long-term issues of clinical merit, societal benefit, economic justification and ethical appropriateness are all significant factors.

Weighed against all the previous discussion we should note that an issue of final affordability must also be considered. There is evidence that societies will invest as much as they reasonably can in the delivery of health care, but that there does come a time when the total cost–benefit ratio of the entire system is not seen as acceptable. At this point forms of 'rationing' begin to occur. Typically such rationing involves the development of standards of clinical practice which select patients for application of technology where the technology is able to do the most good. An example is the restriction placed on the use of artificial kidneys (haemodialysis) in the very old where the net effect of use of the technology, despite its proven capacity to sustain life, is doubtful. As costs continue to rise, examples of this sort seem increasingly likely to emerge and become accepted practice. The problem remains how to manage people's health care aspirations in line with costs—and who in society is to set limits which impinge on ethical and moral sensibilities.

Conclusion

The debate on application of health technology is not new. For some time now the argument that medical technology both increases costs and improves health and longevity has been in the public arena. It has, however, received a major stimulus from the apparently uncontrollable escalation of the costs of the delivery of health services in the last decade. New technologies have undoubtedly had a major impact on the capacity of modern medicine to both alleviate and cure disease and it is clear that many further advances are inevitable over the next decade.

Health costs have ceased to be a medical issue and are now clearly on the political agenda of all industrialised nations in a political process where key stakeholders—government, consumers and providers—are involved. Health care costs are now pitted against the allocation of resources for other social benefits. For example, in Australia, assessment of medical technology is central in the conferences of Australian Health Ministers (AHM), the AHM's Advisory Council, and the deliberations of the National Health and Medical Research Council.

In conclusion, it is clear just how important the careful evaluation of medical technology is and how this may play a part in containing health care costs. Yet beyond this no simple solutions are at hand. Already in the 1980s Australian researchers urged careful consideration of medical technology developments (Bates & Lapsley 1985, pp. 229):

> Both doctors and patients have varied their definition of a 'health problem' to suit the current fashion and the available technologies. This may be an exaggerated version of what has happened, but until very recently, there has been no real challenge to this way of thinking and acting. But the potential consequences of allowing technocrats the indulgence to do anything they can do are quite formidable, and it has become urgent that we should regularly, frequently and publicly discuss how we are going to use our powerful technologies, whether they provide benefits we really want, and whether we really want to go in the direction in which they are leading.

Or, as Wilensky (1990 p. 52), more recently, put it:

> We not only will need to assess how much we are willing to spend on health technology and health care in general, but also how these decisions are to be made and who is to make them. Decisions of this nature will take the best scientific and political minds we have available to strike the proper balance between the costs and benefits that medical technology has to offer.

This chapter aims to foster a better understanding of the evolving debate over medical technology; and, while we acknowledge decisions about such technology are often made on the basis of the non-rational or irrational, we have contributed a suitable framework for consideration of the issues involved.

References

Arney, W. and Bernard, B. 1984, *Medicine and the management of living: taming the last great beast*, University of Chicago Press, Chicago

Bates, E. and Lapsley, H. 1985, *The health machine. The impact of medical technology*, Penguin Books, Melbourne

Bates, E. and Linder-Pelz, S. 1987, *Health care issues*, Allen & Unwin, Sydney

Coile, R. 1990, 'The megatrends—and the backlash', *Healthcare forum*, vol. 33, no. 2, pp. 37–41

Hendee, W.R. 1991, 'Technology assessment in medicine: methods, status and trends', *Medical progress through technology*, vol. 17, no. 2, pp. 69–75

Jecker, N. and Pearlman, R. 1992, 'An ethical framework for rationing health care', *The journal of medicine and philosophy*, vol. 17, no. 1, pp. 79–99

Johansen, K. and Racovean, N. 1991, 'Big ticket health technology: is rational utilisation possible?' *Medical progress through technology*, vol. 17, no. 2, pp. 85–91

Koenig, B. 1988, 'The technological imperative in routine medical practice: the social creation of a 'routine' treatment', eds M. Lock & D. Gordon, *Biomedicine examined*, Kluwer Academic Publishers, Dordrecht, pp. 465–96

Peet, J. 1991, 'Health care: a spreading sickness', *The Economist*, special supplement, July 6th, pp. 3–22

Ragg, M. 1993, 'Heart to heart', *Weekend Australian*, January 23–4, p. 20

Robertson, J. 1991, 'Don't blame medical technology', *Business and health*, October, pp. 109–10

Schwarz, R. 1992, 'The physician and the social contract', *Obstetrics and gynaecology*, vol. 79, no. 1, pp. 1–4

Sullivan, L.W. 1991, 'Health care in America: formidable tasks for the 90s', *Business forum*, vol. 15, no. 4, pp. 10–12

Vimpani, G.V. 1991, 'Resource allocation in contemporary paediatrics: the case against high technology', *Journal of paediatrics and child health*, vol. 27, no. 6, pp. 354–9

Wilensky, G. 1990, 'Technology as culprit and benefactor', *Quarterly review of economics and business*, vol. 30, no. 4, pp. 45–53

Annotated bibliography

Bates, E. and Lapsley, H. 1985, *The health machine. The impact of medical technology*, Penguin Books, Melbourne
Highlights many controversial debates concerned with medical technology—its use, its cost, and its impact on a variety of stakeholders.

Doessel, D. 1988, 'The role of new technologies in rising health expenditures: some empirical results', *Growth*, vol. 36, October, pp. 117–36
Doessel examines the controversy among economists as to whether technologies escalate costs. He provides an outline of ways in which economists have approached the issue, and contrasts Australian and overseas data.

Jennett, B. 1986, *High technology medicine*, Oxford University Press, Oxford
Jennett's grasp of the ambivalence of attitudes to medical technology is refreshing in drawing out the role of the mass media.

Martin, D. and Roseman, C. 1990, 'Health care technology in Australia and New Zealand: contrasts and cooperation', *Health policy*, vol. 14, no. 3, pp. 177–89
An innovative paper that deals with the differences evident in the introduction and control of health care technologies between Australia and New Zealand.

Maynard, A. 1989, 'Is high technology medicine cost effective?', *Physics in medicine and biology*, vol. 34, no. 4, pp. 407–18
Providing an American viewpoint, the author argues that we don't really have a clear picture of whether or not high technology medicine is cost effective because of the absence of adequate economic evaluations.

Rutnam, R. 1991, 'Is equity enough? Feminist perspectives on health technology assessment policy', *Australian feminist studies*, no. 14, Summer, pp. 47–56
Rutman suggests that wider social and ethical impacts of health technologies need to be assessed as well as feasibility, efficacy, effectiveness and economic appraisals. The perspective is feminist and concentrates on equity as a key value and goal in analysing technological implications for women and medicine.

Vimpani, G. 1991, 'Resource allocation in contemporary paediatrics: The case against high technology', *Journal of paediatric and child health*, vol. 27, no. 6, pp. 354–9
This article is an Australian perspective on the issue of high technology in a particularly emotive area of medical treatment—paediatrics. Vimpani explores the contentious issue of the allocation of resources to high technology in comparison to other areas of health care.

Willis, E. 1989, *Medical dominance: the division of labour in Australian health care*, 2nd edn, Allen & Unwin, Sydney
Based on an award winning doctoral thesis, Willis' book argues that the

medical profession dominates a wide range of health care services, controlling knowledge and technology. Not dealing explicitly with iatrogenics—disease caused by the process of diagnosis or treatment—the book provides a background against which such issues can be explored.

PART II

FRAMING THE COMMUNAL

REGULATING TECHNOLOGY
Len Palmer

What is the impact of technology on our lives? Do we as ordinary citizens, workers, students, home makers and consumers make informed decisions on technology choices, needs and means? Do we inhabit a world already formed by technological choices so complex that many seem almost invisible (e.g. the pencil, the telephone, the washing machine)? Do many technologies appear before us as autonomous and beyond choice (computers at work, videos at home, cars to get between them)?

This chapter focuses on how key technology choices are shaped by economic, political and cultural assumptions and systems, often beyond our knowing and way beyond our individual control, although not, in principle, beyond democratic control.

Three basic sets of meanings or questions are implied by the title of this chapter. *What is regulated?* What technologies are subject to regulating? They include communication and information technologies, telecommunications, mass media, reproductive and biotechnology (e.g. *in vitro*), manufacturing (e.g. CAD/CAM), transport etc. Primarily technology means the complex of knowledge, know-how and social practices surrounding machines, tools, processes and products. This social definition of technology is intentionally broad so as to shift our attention away from the mere artefacts of technology, the machines.

How is technology regulated? What do we mean by regulating technology? This includes laws and legislation, government regulations, bureaucratic procedures and requirements, customary practices etc.

Who does the regulating? Organisations at many levels provide contexts in which regulating of technology knowledge and practice occurs: companies, institutions, and lower levels of government at the local level; national states, including the parliament and its machinery, the permanent bureaucracy, the political parties, public servants and their unions; international regulatory bodies, such as the United Nations and its arms [e.g. United Nations Educational Scientific and Cultural Organization (UNESCO), International Telecommunications Union (ITU)], trade arrangements [e.g. General Agreement on Tariffs and Trade (GATT)] etc.

A key understanding in this chapter is that regulating technology centrally involves setting in place legal, political and economic structures which provide part of the context for technological processes and technology developments. To a degree, this understanding focuses on regulating new and future technology developments, but may also concern changing regulatory structures which affect old or pre-existing technology.

Another meaning is also worth picking up from the chapter title. 'Regulating' (the verb) suggests an ongoing process where outcomes are not decided, and also suggests processes of contestation, struggle and possible conflict. 'Regulating' implies a control by external forces or procedures that is not suggested by the term 'regulation' (the noun). In a similar vein, in the area of management and organisation studies, Karl Weick (1969) calls for investigation of processes of 'organising', rather than work on the 'organisation', to reshape the view of the organising process as fluid, dynamic, conflictual and open. The discussion of regulating technology requires such a conceptual setting to do justice to its complexity, its importance for contemporary life, and the power struggles between interested groups implicit in technological processes.

However, before we can begin discussing the regulation of technology a number of crucial issues arising from contemporary social science and the interpretation of contemporary society and cultures must be made explicit. This is to say that the social sciences in general are undergoing dramatic changes that affect what we can say about technology and regulation, as much as what we can say about the family, say, or communication, or nursing or research methods.

Since each of these, the state of the social sciences and the state of technology, are huge subjects this section will focus on some key issues and assumptions about the relationship between technology and social science.

The loss of certainty and objectivity in social science knowledge

Primarily the biggest process at work in the social sciences today is the one questioning their objectivity. In many fields of applied social science and research—and in the heartland of social theory—post-positivist, post-structuralist, postmodernist, interpretivist, feminist and many other critiques have eroded the basis on which we once claimed certainty about what we were studying and saying. Game (1991), for example, argues that sociology produces sociological fictions rather than analysing what 'really' happens in society (1991, pp. 3–5).

Few commentators on technology—and even fewer policy-makers and regulators—have taken this sea-change in the social sciences seriously. Yet it will certainly change the way we perceive technology, policy-making and technological processes and outcomes. One author who has taken this issue to heart is the Scandinavian scholar Cees J. Hamelink (1988). Hamelink suggests that the loss of certainty by what he calls the 'justificationist-model' of rational decision-making leads to a view of technology choice as social gambling. The view that we know what is real, and can justify a path of preferred action, is a kind of pseudo-rationality 'that pretends a privileged access to knowledge and . . . the possibility of preferential judgment based on objective argument' (Hamelink 1988, p. 100).

For Hamelink, we can only know part of what is necessary to make technology choices. To recognise this is to recognise that our technology choices involve a decisive element of gambling social resources for social outcomes.

Let us be clear about Hamelink's argument here. A rationality that Hamelink would accept is one which assumes that knowledge and choices are necessarily shaped by the culture, the period in history, and the people who make those choices. Such people, be they men, policy-makers (usually the same) or entrepreneurs, are influenced by their respective positions, interests and cultures, including assumptions about progress, growth, economic rationality etc. In addition, non-dominant assumptions—the interests of competing groups, ideologies and political strategies—are all part of the process.

Above all, technology choice as social gambling recognises unintended consequences and the inability to guarantee or predict outcomes. It recognises the role of subjective meaning in technology design, introduction and use. It fundamentally assumes that any technology choice is contestable and open to examination. Possible critiques include the digging up of unacknowledged assumptions and implied values, and testing against other values and assumptions. These criteria are also the criteria of good social science in the 1990s.

Three features in good technology choice arise from Hamelink's position. Technology choice must be flexible, have scope for intervention to modify or reverse technological processes if needed, and readiness to learn from errors (Hamelink 1988, pp. 101–3). The dominant models for technology policy, based on belief in certain knowledge and certain outcomes, amount to Russian roulette. Because such policy is unacknowledged gambling, it is gambling without considering the costs of wrong choices.

The costs of social gambling include wasted government research, subsidies and promotion, and lost profitability, productivity and jobs. A good example here is the choosing of so-called 'sunrise' industries based on high-technology developments. Many countries have tried to pick the winner without regard to the indeterminacy of wider international and national contexts for their success. Hamelink's thesis does not lead to avoiding the gamble, but to building in the possibility of the failed gamble. Flexibility, openness to reversals and modification of policy, and learning from errors are the necessary responses to the gamble.

The price of interestedness, uncertainty and partiality in social science

Another feature of contemporary social science that flows directly from the loss of objectivity is the interested nature or partiality of social analysis. By this I mean that any observation or analysis is necessarily partisan, reflecting a point of view, or taken-for-granted assumptions. For example Vig (1988) suggests that there are three main approaches to technology: the instrumentalist, the social determinist and the technological determinist. Each position implies a certain understanding about what makes sense, and what is likely to happen, and rejects or ignores other possibilities. Each is 'interested' and has a stake in technology outcomes. None is impartial, objective or neutral.

An instrumental position assumes that technology is neutral, merely for solving problems (instrumental), often optimistic about outcomes, and usually married to the idea of progress and liberal theories of democracy (Vig 1988, pp. 12–14). Vig cites Mesthene's *Technological change* (1970) as an example of an instrumentalist position. On the other hand a social determinist approach assumes that technology is the result of, and expresses, social priorities and is therefore embedded in the social relations that produced it. It is never neutral. This position involves analysing the social contexts in which technology arises, and identifying the socio-political interests (gender, corporate business, consumer, governmental, bureaucratic interests) that produced the technology (Vig 1988, pp. 14–16). An example is MacKenzie and Wajcman's *The social shaping of technology* (1985). Technological determinist approaches generally share an assumption that technology is autonomous, and is its own driving force (Vig 1988, pp. 16–19). See Ellul's *Technological Society* (1964) for an example.

Technological determinism comes in many forms and is the target of the social determinists in particular, who use their social analysis to expose the power relations behind technological processes and choices. Forester offers an introductory account of technological determinism, where he argues that there is a continuum ranging from soft to hard determinists, with many positions between (1989, pp. 1–3).

My own position is a fusion of aspects of social contextual (or social determinist) and technological determinist approaches. Which is to say that I see technology and technology policy as the result of social forces and interests, but also see technology as changing social relations, practices and culture. This fusion of approaches I place in a focus on the process of technology, technology policy and regulating technology. My assumptions include a view that technology policy both reflects and shapes relations of power and relations of dominance.

The social interestedness of technology analysis applies to policy and regulation as well as to technology itself. These are political issues. A political understanding does not imply that technology necessarily reinforces relations of dominance. In fact the record of regulating technology suggests conflict and negotiation over technology outcomes. Such struggles are implicit in the notion of regulating as a political process. In arguing that technology is not neutral, but the result of contending ideas and interests, I want to underline the way that technology issues are social, cultural and—above all—political issues. From this standpoint the processes of regulating technology are political processes. In a democracy we need to ensure that regulating processes are open, the subject of public debate, and sensitive to public opinion.

Regulation implies both laws and legal statutes, but also the underlying policies and programs upon which the technology gamble is made. Therefore the terms regulation and policy are often used interchangeably in the literature, despite their different legal status. Regulation is the expression of the laws and legislation that arise from the policy process.

Roobeek (1990) suggests, in a survey of the technology policies of seven countries, that three core technologies have become critically important for modern society, and the focus of regulation. They are microelectronics, new materials and biotechnology. Other technology areas like computers, networks, telecommunications and robotics flow from these core technologies. The lack of distinct advantage over any of these technology areas for any country has produced a technology race that no-one can win. Nonetheless, the thinking produced is that of a race mentality, with governments developing policy that they hope will produce competitive advantage over other countries. Policies, especially research incentive programs, are remarkably similar in Roobeek's comparative analysis.

Regulating technology, the terms of the debate

The field of technology regulation reflects processes of power and struggle. Terms used in the debates about regulating technology show the workings of struggle and conflict. For example, the term *monopoly* refers to the dominance of one organisation or company. In Australia, Telecom exercised a government or public monopoly over the telephone network until recently, while AT&T (American Telephone and Telegraph—owned by the Bell Corporation) used to conduct a private monopoly over telecommunications in the United States.

Deregulation strictly refers to the removal of some degree of regulation of technology, implying fewer constraints on competition. In practice, deregulation usually means the breaking of a monopoly through changes to the regulatory environment, and is therefore called re-regulation by many commentators (for example, Westerway 1990, pp. 81–2). The Australian case is as good an example as any here. The entry of the Optus communications consortium into Australian telecommunications, and the letting of contracts to several mobile telephone providers, was permitted by changes to legislation. These changes extended the overall regulatory framework rather than diminishing it, that is, re-regulation not deregulation. On the other hand AOTC has such dominance that its competitor Optus can at best be described as joining an *oligopoly* as a minor player, rather than creating a free market. Similar re-regulation by governments of other telecommunications environments have occurred in the US, Japan and Britain.

The changes in Australia, Japan and Britain also demonstrate the *privatisation* of hitherto public corporations (unlike the US which was historically a private monopoly). The telecommunications providers (monopolies) in most countries were established by their governments, mainly because of the huge costs involved. When legislation is changed to permit privatisation of telecommunications new providers rarely want to supply all telecommunications services in any country because not all services yield a surplus or profit. In fact, one of the justifications of monopolies has been that they have provided standard services (e.g. the telephone) at a basic common price by *cross-subsidisation* of costly services by surplus-producing services (commonly long-distance services).

The supposed advantage of the privatisation of Telecom, and the re-regulation of Australian telecommunications to permit the entry of Optus, is *competition*. The competition between Telecom and Optus is over the high-volume, high-surplus, long-distance services used by business users. The reduction in the prices of long-distance services threatens the cross-subsidy, and is part of the context for debates about introducing (widely unpopular) timed local calls.

The principle of *universal service* embodies the conflicts implied here. Remembering that private companies operate for profit, the worry is that universal service and other issues of access, equity and social justice—commonly called *community service obligations* (CSOs)—will be marginalised. For example, will remotely located, rural and even working-class urban telephone customers be able to afford telephone installation charges and ongoing bills if the cross-subsidy is abandoned? Privatised telecommunications providers might be expected to be interested more in the high-volume, profitable business services than in their CSOs. This is a subject of political debate in all of the countries mentioned where monopolies have been broken up by competition.

Two things are clear from this discussion. Firstly, privatisation, deregulation, re-regulation and the breaking of monopoly do not mean that governments are regulating technology less. Secondly, the last quarter of the twentieth century is a period of dramatic change and reshaping in technology regulation.

Regulating technology at the national level

Most of us think of regulation as a function of the govenment of a nation-state. One influential view of the modern state is that it plays a key role in regulating economic activity. That is, the state routinely acts to limit the capacity of any individual sector, industry or company to reach goals which compete with the interests of the wider industry, sector, or economy. Examples in Australia are limitations on overseas ownership of the media, on cross-media ownership, on mergers and takeovers, and on the environmental pollution by key industries or companies. This might be seen as a negative or limiting role of the state, or it might be seen as a way of including other social values, beside profitability and competition, in the national agenda.

Regulation exists alongside, and is part of the means for achieving, another wider role. This is the task of furnishing a climate in which the economy, sector, industry or company can exist, survive, or succeed. The second role might generally be called infrastructural. It involves creating or maintaining the conditions for production, competition and profit. This might be seen as a positive role of the state, or perhaps providing the basic conditions for the capitalist system to operate. The field of state theory is complex and it is argued that the state is caught in a basic contradiction between its ability to manage these goals, and to generate the funds to carry them out. The state is caught in a 'fiscal crisis' (O'Connor 1973).

Both the regulation and the infrastructural roles are at work in regulating technology. Which is to say that national governments consciously try to improve the conditions for technology production, innovation and success. They also expend energy trying to limit the factors that might impede national technology stability, innovation and competitiveness, as Roobeek (1990) has analysed. Generally speaking, infrastructural activities of states are concerned with securing benefits for the country's economic system as a whole, and technology industries, companies and users as part of this. Regulation activities and policies restrict industries from practices that might damage the wider economic environment. State regulation activities may also include other values like community service obligations, in the *public interest*.

A notable case of deregulation was in Britain, where under the policies of Thatcherist economic monetarism, the monopoly of British Telecom was ended in 1984. This has been a model for deregulation, with Britain called the 'flagship' deregulator of Western Europe (Dyson and Humphreys 1990, pp. 24–5).

In Australia several waves of deregulation have occurred. In 1975 postal services were separated from telephone services, and Telecom created. In 1983 satellite services were introduced as separate from Telecom, with the creation of AUSSAT. In 1992 a second telecommunications carrier, Optus, was permitted to enter into regulated competition with Telecom. A third mobile radio-telephone operator joined in competition with Telecom and Optus from 1993.

In the extensive literature surrounding technology and regulation, almost all commentators believe that regulation is reactive to technology. In fact, crises in regulation policy and wider government policy have occurred in many countries because new technological developments have outstripped the capacity for nation-state regulation to cope. A perfect example here is copyright and computer technology. Chesterman and Lipman (1988) argue that existing copyright laws were based on book printing and ownership, which have been stretched to include sound and music. Legislation in most countries can no longer deal with copyright and ownership issues surrounding computer software and hardware. In addition, even the boundaries between software and hardware are being eroded by new technological developments.

To summarise, a broad trend in many developed countries—especially those espousing free market ideologies—is the breaking up of national monopolies under deregulatory and privatisation initiatives. This has meant a re-regulation and restructuring of technology regulation and policies, rather than a lessening of regulation. Instances of re-regulation in Australian telecommunications show the breaking of Telecom's monopoly, and the entry of new players in the telecommunications game. A new regulatory environment is being shaped. A major social concern in this phase of technological and regulatory restructuring is the survival of policies of access and social justice. The cross-subsidy of services to maintain a universal telephone service is important for equity and access in telecommunications. Conflict at the level of basic social values are part of the political context of regulating technology at the national level.

Regulating technology at the international level

Internationalisation implies the rise of a global economy outside the control of any one country. This powerful reshaping of the world holds the possibility for new contenders for economic or industrial influence to emerge. This is the context for the 'technology race' analysed by Roobeek (1990), noted above. The possibility of creating new technology rankings contains the potential for leading countries, companies and institutions to be challenged by new players.

However, in a survey of European changes to telecommunication and broadcast policies, Dyson and Humphreys suggest that 'whilst considerable destabilization and alteration may have occurred within policy sectors, it is not at all clear that the dominant patterns of relations between these sectors have changed' (1990, p. 30). Dyson and Humphreys' argument contends that the leading technology companies are still the leaders, and the field may be reshaped but is still uneven. The chances for new players in the technology game remain unequal.

Another important finding by Dyson and Humphreys is that there has been a shift in Europe away from a public-service model of communication policies to a market model. Free-market ideologies neglect notions of community service, highlighting instead customer service, usually on a user-pays basis. This has not meant, however, the complete erosion of national commitments to service provision. Public-service providers have on their side large networks, control over programming capability and 'deep roots in national cultures' (Dyson and Humphreys 1990, p. 29).

Two features of the modern world undermine the adequacy of a theoretical approach based only on the nation-state. Firstly, the world has become internationalised in such a way that nation-states are no longer as important as they once were. This has the consequence of diminishing the ability of individual countries to stand alone on issues of trade and regulation, including technology policy. Secondly, the field of technology, especially the many computer and data networks that circle the globe, itself transcends state or country boundaries. The legislation of any one country about technology cannot cover the issues raised, nor regulate such networks. This has led to the rise of importance of supra-national organisations to mediate the interests of nations, corporations and international relationships.

What is required for an adequate grasp of technology regulation processes is the fusion of these two features—internationalisation and transcendence by technology of individual states—into an international perspective that remains sensitive to national cultures and laws. These issues are complex and uneven, revealing the contradictions of trying to achieve global policies from the competing and potentially conflicting interests of nation-states on the one hand, and powerful transnational corporations (TNCs) on the other. Some examples of regulating technology on the international level illustrate such an approach.

Some international 'rules' affecting technology arise from arenas not directly related to technology. For example, the 'Subsidies Agreement' which came out of the Tokyo Round of the 1979 GATT talks set a benchmark for the signatory countries on the effects of government subsidies for technology exports. This is because such subsidies affect profit, technology, industry and government policies—including tariffs—in importing countries. Such technologies include lasers, 64K RAM chips, and the British Concorde aeroplane (Graham 1987, pp. 27–35). Here technology regulation is achieved by the negotiations between countries struggling for survival in, access to, or dominance of, world trade.

Arrangements like the 1979 Subsidies Agreement have implications for world trade law, technological advances and their regulation. They also show some of the complexities involved. The agreement advanced the distinction between technologies that break new ground and those that merely preserve the technological status quo (Graham 1987, p. 35). When the subsidy from the government of a technology-exporting corporation lowers the price of a technology and/or its application to an importing nation, the 'injury' sustained by the importing industry varies according to whether the technology advances technology developments in the importing country, or not. If it does aid technological development, the importing countries and industries may receive benefits to their own industries and development. When subsidised imports compete with local products, the subsidy is a direct cost to the local technology and economy (Graham 1987, pp. 35–41).

We can see in this example that individual countries, their industries and economies, are motivated to reach agreements because lack of international regulation might be worse. Each country defends its interests, and tries to advance its interests, in such an international forum. Major technology companies (often TNCs) are usually involved in such negotiations, either as advisers to their governments, as observers, or as negotiators in their own right.

However, by far the most important international regulatory body with general application to technology issues is the International Telecommunications Union (ITU). The ITU has a 120-year history starting with the emergence of the first communications technologies in the late nineteenth century, and becoming a 'specialised agency' of the United Nations in 1947. Savage (1989) analyses this history and argues that a political perspective is needed to understand the ITU.

The ITU has three general aims: encouraging international cooperation in telecommunications; promoting telecommunications technical facilities; and harmonising international relations over telecommunications (Savage 1989, p. 10). The ITU is recognised internationally in the following three areas: coordination of radio and satellite services; international telecommunications technical standards; and definition and regulation of new and established telecommunications, including computer data flow (Savage 1989, p. 11).

Savage (1989) argues that most accounts treat the ITU as if it were in a vacuum, neglecting vital political contexts for the many decisions made by the ITU. In fact the ITU itself has been the subject of considerable political debate; over whether it has become 'politicised', or as Savage argues, over whether the ITU is necessarily political. Partly the ITU has become a forum for debate of issues around the New World Information and Communication Order (NWICO) raised in the 1970s at UNESCO. The NWICO was a log of demands by third world nations to change the character and direction of news and, later, computer data flow. It was argued that such flows disadvantaged third world countries to the advantage of transnational corporations based in the US and Europe. For their part, the US and Britain argued that the NWICO reforms impinged on freedom of expression, and a free press. The process of reform under the banner of the NWICO, through UNESCO, was undermined by the withdrawal of the US (with its crucial budget vote) and then Britain in the early 1980s. McPhail (1987) shows the impact of this refusal to deal with NWICO issues by powerful first world countries. It effectively crippled UNESCO as an international forum.

While ostensibly dealing with technical issues, like the global allocation of the frequency spectrum for radio, ITU decisions are inevitably political. This is the case both within nations, especially in third world countries, and between nations, especially between first and third world countries—what is often called the North–South divide. The allocation of international satellite and radio frequencies affects the ability of third world countries to maintain, and perhaps control, the information channels that serve their populations.

It should be clear that the politics of telecommunications regulation is part of the ITU and its role in international telecommunications regulation. An important dimension of Savage's analysis shows that the ITU does not merely react to issues raised by technological developments. The ITU has taken on both an ameliorative and a proactive role in advancing political and equity issues raised by third world ITU members, as part of its technical decision-making. To this extent the common view of regulation as lagging technology is inappropriate. The international process of regulating technology is contextualised by the politicisation of the ITU, which appears to be increasing through the 1980s and 1990s.

An important consequence for the widespread recognition of the political nature of regulating technology at the international level is the partisan, or interested, nature of the regulation process. International organisations and forums such as the ITU and GATT talks cannot pretend disinterested or objective discussion. They are predicated on difference in interests, point of view and relative advantage and disadvantage. There is open recognition of partisanship and conflict, and the need for negotiation and compromise. Not that all is sweetness and democratic light in these international settings. The powerful nations and their corporations exercise their power, and the less powerful form blocs to strengthen their own position. The third world have used the need for agreement at the international level to push policies that recognise their comparative disadvantage. There are chances to be taken, gambles to win or lose.

To summarise, regulating technology at the international level involves finding practical solutions to intensely political problems and issues. Nation-states vie as international actors, balancing the need to represent their own self-interest with that of other nation-states in blocs and markets. The domain of international regulation reflects on the one hand the need to find supra-national solutions to problems that confront all countries, and which appear to be out of the hands of each. On the other hand, there is a rising insistence of third world countries for more than technical solutions to access and equity disadvantages; solutions which are sometimes supported by the North. In addition transnational corporations vie as actors in their own right, or as vitally interested parties standing behind the representatives of their countries of location. Trade arrangements such as GATT become hedged by economic and political considerations that directly affect technology export and import relationships. International organisations like the ITU have become increasingly politicised, which is for some a welcome recognition of the political nature of the process of regulating technology at the international level.

There is at the international level an overt acceptance of the inappropriateness of the pretence of objectivity which is usually missing at the national level. A similar recognition at the national level of the political nature of technology policy-making and regulating technology would dramatically improve the quality of those debates. Recognition of different points of view (and their underlying interests) and the openness to debate are fundamental to effective democracy. Too often the participation in technology debates is precluded by the myth of objectivity, the denial of difference and the lack of information for citizens to make useful contributions. Lessons from the international level could improve the national technology regulating process.

Conclusion

In this chapter I have tried to sustain one main argument. I have argued that regulating technology is an irredeemably political process at the national and international levels. Regulating technology is shown to be a contested process, involving struggles over issues of power, dominance and equity. Technology policy and regulatory outcomes are not the result of objective decision-making, but the product of ongoing processes of struggle between states, and between transnational corporations, which have a history into the past and into the future.

Increasingly we live in an international world, with an international economy, an international division of labour, and a concern with international issues which run deep into our national cultures. The interdependence of the national and international levels of our economic survival increasingly marginalises our autonomy as individual states. We must adjust national technology policy to the international context in which it sits. Recognition of difference in needs, interests and values is required at the national level too. Openness to debate must be based on access to information about technological futures, the right to speak, and the development of real possibilities of being heard. Citizens, minorities and lesser companies must form blocs and alliances to build their power and influence. The need for such a political culture in Australia, as in all countries, reaches far beyond the processes of struggle in regulating technology, it reaches into the roots of democracy.

References

Chesterman, J. and Lipman, A. 1988, *The electronic pirates*, Comedia/Routledge, London

Dyson, K. and Humphreys, P. 1990, *The political economy of communications: international and European dimensions*, Routledge, London

Ellul, J. 1964, *Technological society*, Knopf, New York

Forester, T. 1989, *Computers in the human context*, MIT Press, Cambridge, Mass

Game, A. 1991, *Undoing the social: towards a deconstructive sociology*, Open University Press, Milton Keynes

Graham, E.M. 1987, 'World trade law and government subsidies to industrial innovation', *Technology and international relations*, ed. O. Hieronymi, Macmillan, London

Hamelink, C.J. 1988, *The technology gamble. Informatics and public policy: a study of technology choice*, Ablex, Norwood

MacKenzie, D. and Wajcman, J. 1985, *The social shaping of technology: how the refrigerator got its hum*, Open University Press, Milton Keynes

McPhail, T.L. 1987, *Electronic colonialism: the future of international broadcasting and communication*, Sage, Newbury Park

Mesthene, E.G. 1970, *Technological change: its impact on man and society*, Harvard University Press, Cambridge, Mass

O'Connor, J. 1973, *The fiscal crisis of the state*, St. Martins Press, New York

Roobeek, A.J.M. 1990, *Beyond the technology race: an analysis of technology policy in seven industrial countries*, Elsevier, Amsterdam

Savage, J.G. 1989, *The politics of international telecommunications regulation*, Westview Press, Boulder

Vig, N.J. 1988, 'Technology, philosophy and the state: an overview', *Technology and politics*, eds M.E. Kraft & N.J. Vig, Duke University Press, Durham

Weick, K.E. 1969, *The social psychology of organising*, Addison-Wesley, Reading, Mass

Westerway, P. 1990, *Electronic highways: an introduction to telecommunications in the 1990s*, Allen & Unwin, Sydney

Annotated bibliography

Beniger, J. 1986, *The control revolution: technological and economic origins of the information society*, Harvard University Press, Cambridge, Mass
Beniger offers evidence suggesting that contemporary technologies are the latest attempts to respond to a crisis of control triggered by the industrial revolution.

McKay, H. and Gillespie, G. 1992, 'Extending the social shaping of technology approach: ideology and appropriation', *Social studies of science*, vol. 22, pp. 685–716
This article synthesises the social shaping approach (Mackenzie & Wajcman 1985) and the social implications approach, along with insights from other disciplines.

Noble, David 1977, *America by design: science, technology, and the rise of corporate capitalism*, Alfred A. Knopf, New York
Tracing technological developments on several fronts, but especially automated machine control in metalworking shops, Noble shows how America was reshaped by engineers, managers and new forms of work design. A classic.

Reinecke, I. and Schultz, J. 1983, *The phone book*, Penguin, Melbourne
Originally opening up communications regulation to a wider public, this very useful book discusses the issues accessibly without dodging political questions.

Westerway, P. 1990, *Electronic highways: an introduction to telecommunications in the 1990s*, Allen & Unwin, Sydney
Westerway offers an introduction to contemporary telecommunications from an Australian perspective.

Willis, Evan 1988, *Technology and the labour process: Australasian case studies*, Allen & Unwin, Sydney
My chapter in Willis' collection on labour technology is an earlier attempt to deal with technological determinism and users' resistance to technological meanings.

AUSTRALIA'S INFORMATION SOCIETY: CLEVER ENOUGH?

Trevor Barr

Information, as such, is a slippery concept. The term is used to cover so much. Data, copy, intelligence and news—with other specialised connotations—are generally used in narrower senses than the all embracing distinctive term, 'information'. Information can be represented in many different forms and the gathering, evaluating, storing and transmission of information content can involve a wide range of mediating processes. The term, information, is used in such a sweeping manner that economists have struggled to conceptualise the term 'information economy', let alone associated terms such as the 'global information economy'. Little wonder then that the term 'information society'—central to this chapter—is so problematic.

The information sector itself was conceptualised in the 1960s when a Stanford University economist, Marc Porat (1977), re-analysed data in the files of the US Department of Commerce from the perspective of information occupations and information-related activities. This sector not only included the new information industries such as computer workers, but also teachers, lawyers, accountants, clerical workers and professionals. By reclassifying existing occupations into a newly created information sector, Porat was able to argue that the US was already an information society. Over 50 per cent of US employment was concentrated in information activities, and the term 'information sector' now comprises white collar or non-production workers in the labour force.

Technology and the information society

Latterly some commentators have made constructive suggestions for pinning down the information society in terms of its technological implications. A recent parliamentary committee, chaired by Mr Barry Jones, offered a rare definitional attempt (AGPS 1991):

> The information society is a term that has been applied to western, developed nations where communications and computer technology have brought about a concentration of the workforce in the collection, processing and manipulation of data and the organisation of this into information and/or knowledge.

The problem with this definition is that its central notion deals with workforce concentration in information, and that there is little sense of broader social change so often attributed to the widespread introduction of information technology systems and processes. In Porat's day information may have made up the bulk of the US economy, but there is little doubt that the decades since have witnessed substantial social change as the information sector grew to become the information society. The national catch phrase as Australia entered the 1990s reflected the change in social priorities envisioned by local politicians. Australia was to become the 'clever country'. This sounded new and challenging—an acknowledgement that Australia had to move beyond the 'lucky country' ethos and multi-skill up for success in the mid-1990s and beyond.

This positive national image agenda has also been attempted in other countries. The National Computer Board, Singapore (1992, p. 19), sees applications of information technology as the key power behind the future 'intelligent island':

> In our vision, some 15 years from now, Singapore, the Intelligent Island, will be among the first countries in the world with an advanced nationwide information infrastructure. It will interconnect computers in virtually every home, office, school and factory. The computer will evolve into an information appliance, combining the functions of the telephone, computer, TV and more. It will provide a wide range of communication modes and access to services. Text, sound, pictures, video, documents, designs and other forms of media can be transferred and shared through this broadband information infrastructure made up of fibre optic cables reaching to all homes and offices, and a pervasive wireless network working in tandem. The information infrastructure will also permeate our physical infrastructure making mobile telecomputing possible, and our homes, workplaces, airport, seaport and surface transportation systems 'smarter'. A wide range of new infrastructural services, linking government, business and the people, will be created to take advantage of the new broadband and tetherless network technology.

As a useful contrast to Singapore's predictably technological view, the Aspen Group (1989, p. 4) in the USA articulate some social implications in their future visions:

> The explicit purpose and promise of telecommunications in the United States, broadly stated, is to connect individuals and groups to one another and to information and services worldwide, in order to increase economic growth and efficiency and improve the quality of life.

The report goes on to suggest that this would eventually involve change and improvement in the communication process itself:

The Aspen group deliberately uses the verb 'connect' to describe the function of telecommunications rather than the usual 'exchange information'. The group believes that if the full potential of intelligent communications services is achieved, the distinction between linking people through telecommunications and putting them in literal physical contact will blur. If the information society reaches its fullest promise, the telecommunications infrastructure will become not only a helpful substitute for travel, personal meetings, and other expensive means of exchanging information; it will vastly expand the ability of people genuinely to understand each other and their world—beyond what physical presence by itself could ever offer.

Thus the watershed of the 1990s marks a reassessment in a number of countries, with differing cultural traditions, as to the future of their vision of the information society. With specific respect to Australia, it is useful to attempt to state the key trends implicit in the phrase 'Australia as an information society':

- an economic shift away from a prime dependence on traditional resources (raw materials and energy) towards intellectually based value-added goods and services;
- that information, in many forms, constitutes a highly marketable commodity—information goods and services represent an increasing proportion of the gross national product;
- that a substantial number of workers are engaged in productive pursuits that are knowledge generating or information based;
- that the increased pervasiveness of information technology systems and processes has contributed to significant cultural change.

An information society is a society which is characterised by a knowledge explosion, increased use of communications technologies, technological convergence especially between computers and telecommunications, and associated social changes.

The issue of social change is a complex one. Certainly there are social costs associated with the burgeoning information economy. The privileging of information as a key economic market exacts a price from those who supply the information but are unable to control it. Some of these concerns are addressed elsewhere, and they include misuse of information, invasion of privacy and the issues of equity and the public interest.

Paradoxically, too, although Australia is a nation with vast information wealth—compared with many third world countries—many of its citizens are information poor. Ian Reinecke makes the point (1987, pp. 24, 26) that:

selection of information for distribution is determined by its potential to produce profit. No matter how greatly needed, information is seldom supplied to those who most need it but cannot afford to pay for it . . . The net result is that the range of information provided through the new technology in many ways is narrower than it was when the printing press became a universal tool for self expression. Radical, dissident, or even slightly aberrant perspectives are excluded.

The divide between the information rich and the information poor is one of the great social justice issues of the 1990s. Disadvantages are increasingly experienced by the information poor in a world in which information wealth is related to educational opportunity, health, occupation, income, and power. And if the disparity between rich and poor is great within a single society, such as Australia, it is magnified hugely by the disparity between nations.

It is not clear, however, that these disparities, and the technologies which drive them, are an *effect* of the information revolution. There is some argument that the vast increase in information manipulation is the result of a new social dynamic—the need to control processes and information at a distance. This need for control may have been the engine which powered the developments of control technologies which characterise today's information societies.

James Beniger (1986), an American academic, argues that the industrial revolution demanded technologies which would enable manufacturers to coordinate mass-produced goods, which would be efficiently dispersed by mass-transportation systems, to be sold to mass markets. The need to control these processes led to the computerised manufacturing, marketing and supply systems of today. The unifying factor which these processes shared was that they could be expressed in terms of data, manufacturing goals, market penetration ratios and profit margins. By controlling the data, and manipulating the processes to conform to the required outputs, the mass-market industrial processes were also controlled. In effect, the complexity of industrialised society expressed itself in a crisis of control, which continues to have implications in trade-offs between economic developments and the social agenda.

Another American professor, Herbert Schiller (1981), has argued that the processes of the information revolution will continue to accelerate to ensure that new (American) industries arise to enrich or replace the old. Without this technological dynamism, the current trends which allow the massive accumulation of private capital might falter or collapse. The rhetoric which surrounds the expansion of new technologies into fresh arenas tends to be in terms of reducing inequality, allowing greater educational and cultural opportunities, and permitting more national autonomy. The Aspen group would be a good example of such an agenda. Schiller warns, however, that the catalysts for growth of information technology in contemporary societies are privatisation and commercialisation—neither of which is influenced by social need or the larger public interest.

Changing assumptions in the information society

There is no doubt that Australia's information society will be increasingly driven by an ethos of competition. This has resulted in the destruction of the old monopolies, and the creation of a competitive domestic environment with a keen eye to the main chance in the world arena. For Australia is not an information society in splendid isolation. It has to compete internationally with other information societies. Competition is fostered both internally and externally.

Because of AUSSAT's debt there was no viable second carrier within Australia's telecommunications environment which could compete with Telecom without additional financial backing. Optus had its own international infrastructure and could bring this power to bear in its activities in the Australian domestic market resulting from its purchase of AUSSAT. Bell South and Cable and Wireless, each with 24.5 per cent of Optus, brought to the 51 per cent Australian shareholders lines of credit and an international standing which the domestically based Telecom would have been hard put to counter. Telecom and the overseas Telecommunications Corporation were merged to form AOTC—the Australian and Overseas Telecommunications Corporation—to create an Australian-based company with multinational interests. The move was necessary to face the international competition.

Similarly with the deregulation and privatisation of publicly owned airlines: QANTAS and Australian Airlines are merged prior to sale to compete better internationally as offering an integrated service to Australia and within it. Ansett accordingly seeks equivalent links with other international carriers to allow it to compete on the mythical 'level playing field'.

This national/international nexus is only one way in which the assumptions underpinning the information society are changing. Increasingly—whether the focus is newspapers, telecommunications, airlines or media production—the pattern is one of conglomeration of large multinationals into ever larger corporations which are then better able to compete with other huge companies. In this process being big does not guarantee survival, but it is often a prerequisite. The economic-survival dynamic may be pressing from an industry perspective, but it is a poor substitute for democratically discussed social agendas.

In 1985 the problem was expressed like this (Barr 1985, p. 5):

> The information society may well intensify the struggle for survival and strengthen the monopoly of economic activity over the social and political dimensions of our lives. The great danger with contemporary technological

changes and the introduction of new communication technologies is that existing monopolists who own the technologies may strengthen their stranglehold. The most important yet least explored questions about the information society, this new electronic estate, deal with power relationships in society and their effects.

The problem remains. Globalisation is a process from which no-one can opt out.

The ethos of microeconomic reform, then, is leading to the demise of the middle-range company—the nationally sized corporation. These are becoming swallowed up in the larger corporations. It is possible to argue for a future when Australia no longer has a national broadcaster, but where Australians, and the rest of South-East Asia, can tune into the ABC and SBS on the Star TV satellite channel radiated via Hong Kong. Similarly, the commercial channels will no longer be networked Australia-wide from a production centre in Sydney, but will be networked worldwide from production centres in the US. Market-niche, local positioning will be via the topping and tailing of programs and schedules, with news inserts and some local advertising.

Capitalism of the late 1990s may produce a big/small dichotomy, but the small might be beautiful. We may see at the small end of the spectrum an unprecedented blossoming of companies—in media production, security, electronic publishing, design, advertising, self-help, network marketing, and in telecommunications services. By the end of the 1990s most Australians may be employed in companies with a workforce of fewer than ten people.

In this climate, responsivity is something which the government of the day finds especially difficult. The public often do not know the available choices, and are even less likely to be asked their opinion on them. Corporations representing interest groups as diverse as cable, satellite, telecommunications, gas, electricity, broadcasting, narrowcasting, and pay TV are all competing fiercely to get digital highways into the home—and from the home to commercial centres. Yet these decisions are not taken as a result of community debate and wider discussion of preferred options and possible implications. The process appears haphazard and critically dependent upon the most recent financial imperative.

Pay TV, for example, occupied much column space of the inside pages of the heavier dailies over a number of years spanning the late 1980s–90s. Pay TV could be delivered using a variety of mechanisms: by cable—particularly digitally via fibre optics; by satellite; by microwave. The winner of the war to be first to deliver the service was likely to accrue also the benefits of establishing the domestic digital highways network.

From the Australian government's viewpoint, the power it held as the regulator could net financial and infrastructure benefits for the Australian people. On the other hand the new imperatives were competition and choice. Little wonder that the pay TV debate was schizophrenic in its schisms.

The public were framed out of this debate, which occupied much of the early 1990s, by the fast-changing terms of reference and the behind-the-scenes deals. There was also an apparent conviction that, whatever television was offered, it was unlikely to be worth watching at any price. The domestic digital highways, which would have helped the clever country develop a technological infrastructure to rival the intelligent island, went untrodden for years longer than the technological capability—or the commercial sector—would have dictated.

The whole discussion was an example of the fragmentary processes by which vitally important decisions about new communications technology are made in a democracy which does not aspire to technological literacy, or the fostering of community input into technological futures. One of the most astonishing aspects of the pay TV debate in Australia was that it focused so much upon systems—satellite vs fibre optic; Optus vs AOTC; networks vs new players—rather than upon services for which customers would choose to pay.

The debate also characterised the austerity of the 1990s. The exponential growth in telecommunications—both in Australia and abroad, throughout the 1980s—was seen to be an aberration. The scarcity of the resources available to network a continent as large as Australia meant that whichever medium was chosen was likely to become not only the dominant—but probably the only—one for the foreseeable future.

Accountability was also important. In a climate where the public do not own the decisions made on their behalf, they extract a heavy price from governments who have, in the court of public opinion, got it seriously wrong. AUSSAT, which prior to its sale to Optus had become, according to the Telecom unions, 'the lemon in the sky', was a clear indicator that citizens do not need a deep technological understanding to hold strong opinions on new technology issues.

The public interest is a concept almost as slippery as information itself, yet it can only be ignored at a cost. Public interest is the historic ingredient for all open discussions on technological change and development. If something was in the public interest then it needed little else to recommend it—provided that the price was right. But public interest changes with the prevailing philosophies of the government in power. Thus it was for many years in the public interest that Telecom operated a monopoly of telephone services in Australia. This allowed Telecom to cross-subsidise the less profitable areas (untimed local calls, services to remote areas) with the more profitable long-distance and business services, in the wider public interest.

Nowadays the wisdom is that the public interest is best served by competition. The fact that competition is focused on the high volume, long-distance commercial telecommunications market—and that domestic prices for local calls have been steadily rising—does not make it against the public interest. It is naturally in the public interest that Australian business should be permitted to keep its costs as low as possible, to regain international competitiveness and to foster greater national prosperity. Yet is the public interest so pliant a concept as that?

A critical examination of the microeconomic debates reveals the nature of the changing assumptions about public interest in the telecommunications arena. And those who yawn at talk of telecommunications have yet to grasp that, along with computers, they are a fundamental common denominator of today's information society. The telecommunications industry is, by any analysis in the early 1990s, likely to become the largest industry in the world in the next decade. As high technologies continue to develop, so they converge, and the boundaries of telecommunications become increasingly blurred with everything else.

Consider computer networks which use common cables in an open systems environment. Where does the computer element end and the telecommunications element begin? Or a service which enhances a transmission line (with data-switching or processing capabilities)—is that a traditional telecommunications activity or a new computer offering? Or the digitisation of information itself—sound, images, video, text, data etc; where would the Integrated Switched Digital Network (ISDN) be without its telecommunications infrastructure—or its computer interface? So discussions about telecommunications are highly relevant to Australia's information society, and to the public interest.

Often the public interest is piggy-backed upon private commercial interest. Thus in the old, monopolistic days, when Telecom's local sourcing proved so profitable for local subsidiaries of transnational corporations, it was the Australian subsidiaries of Alcatel and Ericsson which particularly benefited from the local sourcing arrangement. With Optus in competition, DEC (Digital Equipment Corporation), Nortel Australia, Nokia Telecommunications of Finland and Fujitsu all earn a larger slice of the action. The experiences of the UK and New Zealand, however, show that deregulation practices can decimate a significant sector of the local industry.

David White (1991), the Victorian Minister for Manufacturing and Industry Development in the Kirner administration, summarised some of the lessons of history:

> Twenty years ago, Australia had a significant R&D presence in computer development, leading the world in some areas.
>
> Today, in computer hardware manufacturing, there are some isolated success stories (such as IBM in Wangaratta) and we still have a core of highly

skilled software developers, but we do not have a significant computer hardware manufacturing sector despite the enormous growth of this industry world-wide.

Contrast this with telecommunications equipment, where major multinationals such as Ericsson Australia, Siemens and NEC (Nippon Electric Company) Australia have substantial manufacturing operations and now have significant exports from this country.

In less than twenty years employment in electronic manufacturing as a whole has halved, while in telecommunications equipment it has not declined. The difference is Telecom.

David White was referring at that time to Telecom's government-regulated policy of purchasing local equipment. This was a major stimulant for the growth of locally based industry. And there is no suggestion that Telecom was propping up uncompetitive industries. The telecommunications industry was one of only ten businesses in Australia, identified by the Australian Manufacturing Council, which generated more than $100 million in exports in 1991.

In addition to the local subsidiaries of the transnationals (Ericsson and Alcatel earned $110 million in exports between them in 1990–91), twelve indigenous companies contributed $54 million to Australia's export revenues: AWA (Amalgamated Wireless Australasia) Ltd, Datacraft, Dataplex, Exicom, JN Almgren, Megadata, Mitec, Netcomm, Olex Cables, QPSX (Queued Packet Switched Exchange) Communications, Scitec Communication Systems and Uni-Lab Telecommunications. The total telecommunications manufacturing export earnings for 1990–91 were $250 million—with a government target of $2 billion by 1997.

The concern, however, in the deregulated environment of competition and choice, is that the rules of telecommunications equipment purchasing which were central to the growth of the industry in Australia have been forgotten. Local development of telecommunications technologies is directly attributed by the manufacturers to the effects of the government-regulated local sourcing policy upon Telecom. Whether AOTC and Optus will choose to follow suit, and whether the government should regulate that choice, are highly political questions. The public interest remains a key concept for the debate.

Whither—or wither—the public interest?

In [public interest] debates, two primary perspectives require representation, but in most cases are absent. One is the perspective of those groups in society that may be significantly affected by the policies adopted, but which do not have a sufficiently organised financial vested interest to mount a representation, e.g. users of the public telephone service, children's interests in television or probable victims of technological change. This perspective is necessary to ensure that in the final balancing of interests underlying most policy decisions, the interests of important public segments are not omitted.

The second perspective is that of society as a whole, focusing directly on the overall structure of benefits, costs and consequences for society. This would include an evaluation of economic externality, public good, social and cultural consequences of policy options (Melody 1990, p. 16).

The Australian Broadcasting Authority (ABA), which replaced the Australian Broadcasting Tribunal in 1992, is part of the new-look, new legislative objectives for the mid–late 1990s. *The Broadcasting Services Act 1992*, which heralded the change, marks the transition 'from public interest to market principles of regulation' (Davies & Spurgeon 1992, p. 85). Which is not to say that the Act ignores issues of public interest: it simply relegates them to a secondary importance. For example, the Act calls upon the Australian Broadcasting Authority to regulate broadcasting in a manner which (s4(2)(a)): 'enables public interest considerations to be addressed in a way that does not impose unnecessary financial or administrative burdens on providers of broadcasting services'.

The ethos of the Act is to allow market forces the primary role in regulating the industry. Much that was previously regulated 'in the public interest'—such as Australian music quotas on radio, or the quality of broadcast signals—is now left to self-regulation by the industry, or is deregulated entirely. This is a technological environment in which the broadcasting spectrum is no longer regarded as a scarce resource. It has been stretched to virtual infinity by the prospect of cable and satellite broadcasting, and by developments in digital formats and transmission systems which will greatly increase the efficient use of the free-to-air broadcast spectrum.

A licence to broadcast is no longer a privilege which should only be entrusted to worthy companies. Allocation of broadcasting rights by the ABA is on the basis of 'the ability to pay for broadcasting spectrum access . . . [not the] public interest criteria . . . such as the financial, management and technical capabilities of an applicant to provide a service of reasonable quality' (Davies & Spurgeon 1992, p. 87). Nor is there any provision for the public to have a right of involvement in the licensing and programming decision-making of the ABA. One of the key points at which the public were able to express their views, the licence renewal enquiry, has been removed. (Broadcasting licences are now automatically renewed.) Instead, the Act makes provision for many decisions to be made routinely, or in private.

If Melody's vision of public interest (above) is accepted, then there is increasingly little room for it in the legislative framework of Australia's information society. Instead, the public interest—the interests of those without the power to organise on their own behalf—is now replaced by the profit motive. The only public interest likely to be served is that which promises a return upon cash invested. Ian Reinecke's (1987, p. 38) manifesto for an equitable information society appears increasingly out of step with the new imperatives. Yet it may be more needed than ever:

> In every area of information distribution, the disinterested process of providing information to those who most need it must be pursued if we are to live in a democratic society . . . Public resources which would enable libraries, government departments and other storehouses of information to provide society with what it needs to know should be injected with funds. Radio broadcasters who seek to serve community needs before shareholders' demands for return on investment should be eligible for public grants and loans. Public television should be given more finance, political support, and an obligation to inform rather than simply entertain.

Politics not policies

The future for a whole-hearted recognition of public interest concerns within Australia's information society looks bleak. An analysis of the processes by which Australia's telecommunications manufacturing industry stands threatened, by which the debate upon pay TV concentrated on systems rather than services, and by which the public interest became a luxury which should not 'impose unnecessary financial and administrative burdens', suggests the following: a small group of policy-makers holds the balance of power in the key decision-making process. This group is composed temporarily of a few politicians who are members of the government of the day, together with their key advisers, and the more senior permanent members of the bureaucracy, with selected inputs from commercial interests and from effective pressure groups. There is little place in these processes for public debate or a wider consideration of the social agenda.

Technology and communications policy in Australia has been characterised for decades by an appalling neglect of social philosophy. Governments have lacked an integrated frame of reference for decision-making in the communications and wider technological fields, based on thorough social investigation, strategic planning, strong public awareness and a careful assessment of options for the future.

Ad hoc decision-making is the inevitable consequence of an essentially irrational political process, from which clear comprehensive policy statements fail to emerge as blueprints for action. The response to technological choices by governments in Australian communications policy has essentially been either a series of short-term pragmatic solutions—which may be one way of looking at AOTC—or the shelving of decisions in complex policy areas, such as pay TV. Inevitably, policy errors or neglect of the past must be paid for by the community as a whole. There is a tendency in the communications field for bad policy decisions on major issues to establish their own ratchet effect—decisions take place on a limited range of options which then permit motion in one direction only.

The overall position of technology and communications policy is a chaotic frame of reference, with a generally ill-informed community, and a political process ill-equipped to come to terms with major issues for the future. A prime paradox of public policy is that major policy decisions are made in forums which are essentially private and closed, and within institutions able to wrap themselves in secrecy. In the fields of technology and communications, there is no permanently established mechanism whereby a wider cross-section of the public can effectively offer an input into the policy/decision-making process; and later effectively analyse the outcomes.

Moreover, the mass media are structured in such a way that relatively few people have access, and the accepted conventions of presentation restrict discussion and exploration of complex subjects. In the context of debate about technological change and the future, the political system as a whole has the effect of inhibiting curiosity and limiting public awareness. In terms of the knowledge and confidence to contribute to discussions about technological futures, the Australian people are very information poor indeed. In this climate the public learns passivity, helplessness and cynicism in the face of technological choice.

Conclusion

It is impossible to trade-off social agendas against economic development without compromising an increasingly untenable notion of the public interest. The costs paid by Australia for its information society include misuse of information, invasions of privacy and increasing inequity as the information poor lag further and further behind the information rich.

The assumptions which currently drive Australia's information society are centred upon microeconomic reform—which is to say the twin gods of competition and choice. Competition is not an end in itself, however, and there has been too little attention paid to the defining of policy objectives which competition should be helping the nation to achieve. Without a sense of the wider public good, and public participation in the discussion, the deregulated communication and information industries are at risk of alienating the Australian people further, and turning what really could be a clever country into one of cynicism and resentment.

What is required is an inclusive agenda for public discussion upon technological futures, and open debates about the costs and benefits of technological change. If the public is to be interested in such issues then the concept of public interest requires more than lip service. The slippery redefinition of the public interest which states that what is good for the market is good for us all could prove, in the long run, to mark the end of Australia as a democratic information society.

A democracy needs to be informed, and to have a right to information. It also requires extensive public debate and widespread participation in decision-making. These are policies which should overarch narrow definitions of politics. They constitute a necessary social agenda if today's information societies are to be communities in which people (rather than data) thrive.

References

AGPS 1991, *Australia as an information society, grasping new paradigms*, Australian Government Publishing Service, Canberra, May

Aspen Group 1989, *Statement of goals and strategies for state telecommunications regulation*, The Aspen Institute, USA

Barr, Trevor 1985, *The electronic estate: new communications media and Australia*, Penguin, Ringwood, Vic

Beniger, James 1986, *The control revolution: technological and economic origins of the information society*, Harvard University Press, Cambridge, Mass

Davies, Anne and Spurgeon, Christina 1992, 'The Broadcasting Services Act: a recognition of public interest and market principles of regulation?' *Media information Australia*, no. 66, November, pp. 85–92

Melody, William H. 1990, 'The information in IT: "Where lies the public interest?"' *Intermedia*, vol. 18, no. 3, June–July, pp. 10–18

National Computer Board, Singapore 1992, *A vision of an intelligent island: the IT 2000 report*, National Computer Board, Singapore

Porat, Marc U. 1977, *The information economy, definition and measurement*, Office of Telecommunications, US Department of Commerce, Washington DC

Reinecke, Ian 1987, 'Information and the poverty of technology.' *Communications and the media in Australia*, eds Ted Wheelright & Ken Buckley, Allen & Unwin, Sydney

Schiller, Herbert 1981, *Who knows? Information in the age of the Fortune 500*, Ablex, New York

White, David 1991, *Towards international best practice in telecommunications—preparing Australian industry to compete in global markets*, Opening address, CIRCIT, Melbourne, 20 September

Annotated bibliography

Barr, Trevor 1985, *The electronic estate: New communications media and Australia*, Penguin, Ringwood, Vic

A complex series of interactions and interrelations has developed between electronic, communications and telecommunications technologies which together constitute the electronic estate. Written before AUSSAT went into service, this book reflects and contributed to debates about the information society in a satellite era. The discussion remains of relevance in its consideration of policy priorities, public interest and issues of equity.

Melody, William H. 1990, 'The information in I.T.: "Where lies the public interest?"', *Intermedia*, vol. 18, no. 3, June–July, pp. 10–18

This short and very readable article is the ideal introduction for people who may not be certain that they understand what 'the public interest' really is.

UNIVERSAL SUFFRAGE? TECHNOLOGY AND DEMOCRACY

Julianne Schultz

In the not so very distant past, before computers became an ordinary element of everyday life, infiltrated into homes, cars, offices and factories, popular images of a possible technological future revolved around two diametrically opposed scenarios.

The first helped to foster a genre which found life in Kurt Vonnegut's *Player piano* and in many other books and films. In these, technology was an invisible big brother controlling an enervated, automated and downtrodden population. In the alternative scenario, conjured independently both by those who had experienced the possibilities of participatory politics in the 1960s, and by the nascent high-tech corporate public relations departments, technology provided liberation from drudgery and ignorance. Just as life in the Athenian *agora* with its participatory democracy had depended on the invisible but material base of slavery, so technology would be the new slave, freeing more time for involvement in a democracy driven by participation, not consumption. In technological utopia the attitudes and wishes of each individual could be electronically considered—a society in which every individual's whim could be taken into account. Technology could create the ultimately refined participatory democracy.

Like all fantasies, these scenarios contained seeds of truth. The rapid technological change which has engulfed the world in the last decades of the twentieth century, however, has been both more banal and more far-reaching in its consequences than proponents of either fantasy could have anticipated. Microelectronic technology, as much and possibly more than other technological breakthroughs (printing, mechanisation), has consequences for every aspect of the lives of those in the advanced world; while leaving those in the less developed world grappling with systems which to western eyes seem frustratingly inefficient.

It should not be surprising that major developments in science and their application should impact on the fabric of political and social life. It has long been the case. As Thomas Carlyle wrote in 1836: 'He who first shortened the labour of copyists by device of moveable types was disbanding hired armies, and cashiering kings and senates and creating a whole new democratic world: he had invented the art of printing' (cited in Keane 1991, p. 164). It was, however, several centuries after Gutenberg's invention, and only following another series of inventions, that liberal democracy became a recognisable political form. The Industrial Revolution—and the *bourgeoisie* created with the wealth generated by new methods of production and distribution—was the crucial trigger in developing a liberal democracy in the nineteenth century (Garnham 1990, p. 106). Decisions made in the next decade, as microelectronic technology is applied to an ever widening range of uses, will determine whether the technology fosters a new set of political and social relations as significant as the development of liberal democracy.

At its most banal microelectronic technology is simply a tool which aids the efficiency with which tasks can be completed. At its most far-reaching it has the capacity to change what we know, what we believe in, what we do, how we do it and even who we are. Microelectronic adaptability— especially when applied in a deregulated and market-driven environment—has the capacity to affect the ability of citizens to function in a democratic society. Consequently it has the potential to influence the very nature of the democracy itself.

That technology should be considered in terms of its potential to impact on democratic systems is itself interesting. Nonetheless, the relationship between technology and democracy is not simply confined to its impact on the political and social processes inherent in modern democratic societies. It may be possible to apply principles of openness and participation to the design and use of microelectronic systems themselves. This potential is, however, unlikely to be generated by the corporations controlling the technology, unless they can identify a market in such design, or are directed to incorporate democratic principles by supervising or implementing agencies (Clarke 1992).

The speed and scale of change in microelectronic technology has been a fundamental component of the campaign for a less regulated, more market-driven economic system. The technology has been used as a stalking horse to test the old verities: television no longer needs to be regulated because of scarcity, there is the potential to deliver tens of channels; telecommunications networks are no longer just public infrastructure, but the building blocks for economic growth and transformation; efficiency is more important than *bourgeois* rights such as privacy. In country after country the argument has been convincingly put that the fetters of regulation needed to be removed for the transformative capacity of the technology to be realised.

Conceptualising democracy

Technological deregulation in the western world coincided with the end of the cold war, a resurgence of democratic rhetoric and an increasing interest in ways of conceptualising—and realising the potential of—democratic systems. As a result of a coincidence of timing, if nothing else, technology and democracy are inextricably linked. For the first time in many decades discussion about the nature of democracy is moving onto the political agenda in many western countries.

As a number of writers have pointed out, the struggle towards democracy is long and incomplete. Three of the guiding principles in this movement have been the desire to establish a check on arbitrary rules, to replace the arbitrary rulers with just and rational ones and to enable the population to participate in the making of rules. Modern ideas of freedom and toleration, stemming from John Locke in the late seventeenth century, are founded upon religious toleration. The overriding right to freedom of conscience in which no government has the right to tamper carries with it other rights, including those of freedom of opinion and expression.

Although visions and theories of democracy have recurred since Plato and Aristotle, the main western tradition was, until relatively recently, anti-democratic. The promise of the utopian, democratic and classless society was incompatible with the hierarchical reality. The practice of liberal democracy has only been a reality since the early nineteenth century, when 'theorists found reasons for believing that "one man one vote" would not be dangerous to property or to the continuance of class-divided societies' (Macpherson 1977, p. 10).

In his lively book *The life and times of liberal democracy* Macpherson (1977) outlines three key phases in the development of democracy: protective, developmental and equilibrium models. The protective model, articulated by Jeremy Bentham and James Mill, which has proved to be both perceptive and resilient, contains the overriding principle that:

> man [*sic*] is an infinite consumer, that his overriding motivation is to maximise the flow of satisfactions to himself from society, and that a national society is simply a collection of such individuals. Responsible government . . . was needed for the protection of individuals and the promotion of the Gross National Product and nothing more (Macpherson 1977, p. 43).

John Stuart Mill—son of James Mill—put forward a developmental model which included a moral element lacking in the protective model. Democratic government was shaped by a desire to improve and develop each individual's capacities for the good of society, while at the same time expecting individuals to act as maximising consumers and appropriators.

Techniques of economic analysis are applied to political systems in the equilibrium model, which largely prevails today. Democracy is seen as a mechanism for choosing and authorising government, more than a description of a kind of society, or a set of moral ends (Macpherson 1977, p. 78). This model, first outlined by Joseph Schumpeter in 1942, is devoid of moral content and treats democracy as a market mechanism in which the voters are consumers and the politicians the entrepreneurs.

Macpherson, enthused with the potential of participatory politics and technology, then casts a fourth model. This he describes as a new participatory form of democracy in which market assumptions about human nature would be abandoned, and the maximising consumer would be replaced by an informed and participating citizen in a society with greatly reduced economic and social inequality.

The idea that recent and expected advances in computer technology and telecommunications will make it possible to achieve direct democracy at the required million-fold level is attractive not only to technologists, but also to social theorists and political philosophers (Macpherson 1977, p. 95). This potential remains, but the market has proved more resilient, coercive and adaptable than Macpherson anticipated in the mid-1970s.

John Keane, director of the Centre for the Study of Democracy at the Polytechnic of Central London, has written extensively about democracy, a concept which he believes is presently dogged by confusion:

> The concept of democracy is not infinitely elastic, even though its principles have been interpreted in diverse ways, as their custodianship has changed hands. The struggle to control the definition of democracy is an intrinsic feature of modern societies. And yet democracy is not a word which can be made to mean whatever we choose it to mean. At a minimum . . . democratic procedures include equal and universal adult suffrage; majority rule and guarantees of minority rights, which ensure that collective decisions are approved by a substantial number of those entitled to make them; the rule of law; and constitutional guarantees of freedom of assembly and expression and other liberties, which help ensure that the people expected to decide or to elect those who decide can choose among real alternatives . . . In large-scale, complex societies regular assemblies 'of the people' as a whole are technically impossible. Direct democracy, the participation of citizens in the agora, is suited only to small states and organisations in which the people find it easy to meet and in which every citizen can easily get to know all the others (Keane 1991, p. 169).

Interactivity and the profit motive

Within this framework it is easy to see ways in which technology has the capacity to influence the democratic fabric of society. All too often in a market-driven economy the rights of citizens are equated with the rights of consumers, and the exercise of citizenship is seen as just another consumer choice. This has been reinforced by developments like interactive television, which build upon previous experience of phone-in polls. The most likely use for interactive television in Australia is audience response to game shows, and snap polls on current issues (should the convicted murderer be executed?). In the United States, Bill Clinton's contemporary simulation of the agora came shortly before his inauguration when he held a nationally televised summit of leaders incorporating a 'talk back' component—any citizen with a phone and a television could contribute to the discussion. The illusion of participation was powerful and compelling. This and similar uses of the technology highlight critical questions about technological participation in democracy: who decides who can participate and who formulates the questions? (Macpherson 1977, p. 95)

The promise of democratic interactivity has been greater than its reality in almost every country. The combination of a search for profit, deregulation and reluctant governments has left the development of the technology, and its potentially immense implications for citizen feedback, in the hands of the market. Prestel in Britain, Telidon in Canada and Bildschimtext in Germany have been available only on a 'rather user-unfriendly and closed user group basis because no private investors, advertisers or governments have been willing to shoulder the risks associated with its introduction for wider public use' (Keane 1991, p. 75).

In France by contrast a graphic-based videotex system, Minitel, was made available free to householders through the initiative of the French telecommunications, Direction Générale des Télécommunications (DGT), which was keen to establish a mass market for the service. More than five million terminals have been installed. The system contains 12 400 service codes and is used by millions of French citizens to search a telephone directory, reserve a ticket, teleshop, bank, learn a foreign language, receive news and send mail. Although Minitel is not without its problems—and is used least by the new underclass at whom it was originally targeted—the success of the French system has demonstrated the limitations of a commercially driven approach to the introduction of technologies.

Given the *realpolitik* of our age, the reluctance of governments to explore the potential of technology to create a participatory democracy is scarcely surprising. Even so, that potential should not be underestimated. When left solely to commercial developers, the potential inherent in new technology will be realised in ways which maximise profit, rather than maximising democratic involvement. Further, the danger is that the price of access will exclude many people:

> We would find it strange now if we made voting rights dependent upon purchasing power or property rights; yet access to the mass media, as both channels of information and forums of debate, is largely controlled by just such power and rights (Garnham 1990, p. 111).

When considering the implications of technology for democracy the most obvious area of concern is the capacity of the technology for surveillance of the thousands of individual decisions of individual citizens at work, in their dealings with government agencies and in their homes. See Roger Clarke's chapter, 'Dataveillance: delivering *1984' for detailed discussion of this issue*. One of the crucial political adaptations of information technology is the refinement of direct mail procedures for political campaigning. Gone are the days of blitzing all members of an electorate with identical materials. The computer matching of interests, concerns, ages, addresses, incomes and political preferences has been refined to such a degree that voters receive differentiated information. Those people concerned about environmental issues can be targeted by political candidates using one communication, while their neighbour's concerns about employment can be identified and addressed differently by the same candidate. The potential for abuse of the political system, entrenching the citizen as the passive consumer, is clearly anti-democratic. Active democracies empower citizens to play a part in setting the political agenda. Significantly, when it was proposed that one way to reduce the electoral gerrymander in Queensland would be by providing voters in the more remote parts of the State with greater access to facsimile and telephones, the Electoral and Administrative Review Commission preferred to continue weighting votes.

Media, technology and democracy

The decisions that voters make are influenced by the information available to them through the media. Arguably the media have been more profoundly influenced by the introduction of microelectronic technology than any other sector. As a result of changes in production, distribution and reception technologies the media industry has changed shape and direction in the last twenty years. It has become one of the most profitable industries and its capacity for political influence remains unrivalled. Both directly and indirectly the technology employed by the media has the capacity to influence the public agenda. This has been recognised by both its critics and its proponents. In 1983, well before the world watched with horror as the tanks rolled into Tiananmen Square, or with amazement as the troops landed in Somalia and the smart bombs exploded in Iraq, Arthur C. Clarke—one of the most important individuals in conceptualising and developing the technology which has enabled the spread of global broadcasting—said:

> The very existence of new information channels, operating in real time and across all frontiers, will be a powerful influence for civilised behaviour. If you are arranging a massacre, it will be useless to shoot the cameraman who has so inconveniently appeared on the scene. His picture will already be safe in the studio five thousand miles away and his final image may hang you (Shawcross 1992, p. 242).

The former communist regimes of eastern Europe found blocking media from the west increasingly difficult, and the application of even limited electronic technologies made the production of the *samizdat* publications of the Opposition much easier (Schultz 1990). In Panama, as President Noriega struggled to maintain power, he shut down the country's independent media. A Panamanian exile managed to produce an alternative newsletter and sent it to facsimile receivers in banks, law offices and travel agencies throughout the country. Within hours up to 30 000 photocopies of the paper were on Panama's streets (Litchenberg 1991, p. 372). The use of the technology of television broadcasting by indigenous communities in Australia and Canada has fostered new and democratic expressions, quite at odds with the original and mainstream application (Michaels 1986).

Although microelectronic technology has the potential to influence and shape democratic activities, it has been developed and applied without heed for democratic ambitions. Occasionally adapted for use in a more open and responsive manner, new communication technologies have been devised by, and for the use of, transnational corporations with global activities and a need for instant communication and access to data. Hamelink notes that information technology has not been promoted to meet basic needs, but to support the expansion of transnational capital.

Studies on the deployment of telephony, educational television, and satellite communication suggest that while peripheral peoples do benefit from the introduction of such technologies, the primary beneficiaries are the foreign and national elites; frequently also, the intended development objectives are not achieved, and serious balance of payments problems occur, as the hardware has to be paid for regardless (Hamelink 1990, p. 225).

This pattern is repeated in sector after sector, but the consequences are particularly far-reaching in the media, whose product is the information which helps shape the values, attitudes and judgments of citizens. This sector has evolved into a global business dominated by a few corporations. The consequences of this pattern of ownership for individuals and governments are considerable, and go to the heart of the discussion about the relationship between technology and democracy. The development and application of the technological tools of this generation have broadened the base and range of media activities possible from newspapers, magazines, books, radio, television and film, to include databases and interactive information systems, pay, cable, satellite and interactive television. With each new development the potential for profit to be taken from the packaging and sale of infor-mation is increased.

Examining the media is particularly instructive when considering the relationship between technology and democracy. Traditionally the media have claimed a special role as the protector of democracy—the upholder of the fourth estate. Media executives and workers like to think of their industry as the one institution with the capacity to watch out for the interests of citizens and to provide a check on the parliament, the executive and the judiciary. Although the media have become essentially commercial, their vociferous objection to regulation is still couched in terms of the need for the information sector to be untrammelled by government oversight, in the interests of the effective functioning of a democratic society. By equating the public interest with the information industry's self-interest the possibilities of democratic control have been inhibited.

The notion of the media fulfilling the role of the fourth estate has been a remarkably resilient concept, almost endlessly capable of reinvention to accommodate changing technologies, regulatory environments and political priorities (Schultz 1992b). Now, at a time of widespread faith in deregulation, media companies may begin to claim that they are just another business after all and that no special rules need to apply. If this occurs a much more fulsome acknowledgement of the primacy of the entertainment function of the media is likely.

Media diversity and the public interest

The tension between the public service and the entertainment functions of the media has been resolved in different ways in the print and electronic sectors, but in both cases the equation of the public interest with commercial self-interest has been ruthlessly made.

In the electronic media the argument about the need for fairness and public service was largely predicated on the assumption of the scarcity of resources (the limited spectrum), and the need for its careful allocation underpinned several generations of regulatory activity. Further, those concerned about the effects of television on the community helped foster a research industry whose findings have tended to support the regulatory model (Cunningham 1992). Now that the scarcity argument has broken down with the development of additional means of delivery self-regulation has been adopted. The effects research has been jettisoned and, at the same time, the public broadcasting sector has been placed under severe financial constraints. The public service and democratic functions of the electronic media have been left to find a place for themselves in the commercial marketplace.

Concern about the democratic role of the press has most frequently found its expression in public inquiries, notably: the Hutchins Inquiry in the United States in 1948, five British and Canadian Royal Commissions, the Norris Royal Commission and the Lee Parliamentary Inquiry in Australia. These inquiries reiterated, with varying degrees of insight and perceptiveness, analyses of the role of the press which incorporated statements about the centrality of the medium to the effective functioning of a democratic state. In the background—often acting as the trigger for the inquiry—has been the dramatic reduction in the number of newspapers published in each of these countries. This has been coupled with a reduction in the diversity of opinion expressed within print media, declining circulation figures and increasing public disdain. As a result the democratic role of the press in providing information and analysis has been circumscribed.

To some extent this decline in the number and diversity of newspapers has been offset in the eye of the regulators and the politicians by technologically facilitated growth in the range and diversity of other print outlets. The Lee Inquiry (1992) conceded that the closure of nineteen newspapers in less than a decade may have had an impact on the diversity of information available in Australia, but argued that future diversity would be ensured by new technology. The committee concluded that these technological developments would ensure that the democratic responsibilities of the print media would be fulfilled (Schultz 1992a).

This analysis minimises the fact that new developments in the print industry are directed by a few companies, often transnational corporations with established dominance over existing media. Ben Bagdikian (1990) points out that over seven years the number of companies controlling most of the media production in the United States went from fifty to ten. On a worldwide scale the predictions are that fewer than ten companies will dominate the entire global production of news, information and popular culture by the turn of the century. If information is the key to the effective functioning of a democratic society then the prospect of information being commercially controlled by such a small number of giant corporations is of immense concern—their power and scale puts them above the democratic system.

At the same time that control of the media is being concentrated in a few corporations, the amount of information produced has increased exponentially. This raises questions about the capacity of citizens to deal with information overload—an avalanche which threatens to engulf them. More information is available than ever before, but the capacity of individuals to process, sort, select and use appropriate information may have been diminished. As Keane comments, 'The world seems so full of information that what is scarce is citizens' capacities to make sense of it' (Keane 1991, p. 183). The notion of the information grazer browsing through available stocks of data, but barely able to make coherent sense of it beyond the personal gratification it delivers, is articulated by Neil Postman in *Amusing ourselves to death* (1984). It may become a recurring theme.

One of the problems with the information avalanche is that those with the greatest need for information tend to have the fewest skills in accessing and using it. Another problem is that of information pollution, how to deal with inaccurate, malicious and 'just wrong' information. Melody notes a need to 'develop policies for managing information quality to ensure that a surplus of erroneous and misleading information does not debase the value of information . . . Ignorance based on misinformation will be the pollution of the information age. Overcoming it will be a formidable challenge' (Melody 1990, p. 18).

When this concern is combined with the market-driven forces which predominate in public policy discussions today the potential challenge to the fabric of democratic society is apparent:

> In democratic societies the scope and meaning of liberty of the press and the process of representation will always be contentious, whereas a society that is drugged on either money or political authority and which contains no controversies over freedom of expression and representation, is surely a society that is dying, or dead.
>
> It may be that the US, Britain, Germany, Italy and other countries are in the process of becoming such societies . . . it may be that too few people will appreciate that the current reorganisation of the media will do more over time to shape irreversibly the future of our societies than all the rest of deregulation put together.
>
> It may be that citizens will no longer invest any hopes in public life. Perhaps they will amuse themselves to death . . . perhaps they will be persuaded to privatise themselves to regard politics as a nuisance, to transform themselves silently and unprotestingly from citizens to mobile and private consumers. Perhaps they will forget that the media of any society are among the most important institutions, and that the courage and independence they display are always a register of the state of morale and vigor of other bodies from schools, trade unions and churches to legislatures, governments and courts of justice (Keane 1991, p. 193).

Conclusion

The question of 'where lies the public interest?' is a difficult one. It is the essential element, however, when considering the relationship between the media and democracy—especially in an age of technological plenty:

> Public communication lies at the heart of the democratic process . . . citizens require, if their equal access to the vote is to have any substantive meaning, equal access also to sources of information and equal opportunities to participate in the debates from which political decisions rightly flow (Garnham 1990, p. 104).

The decisions that are made in the next decade in the design, implementation and regulation of technology will determine whether microelectronic technology will enhance liberal democracy, or trigger even more widespread disaffection with public institutions.

References

Bagdikian, Ben 1990, *The media monopoly*, 4th edn, Beacon Press, Boston

Clarke, Roger 1992, *Balancing benefits against risks in the proposed health communications network*, keynote address, seminar of Australian Medical Informatics Association, Perth, 24 October

Cunningham, Stuart 1992, *Framing culture*, Allen & Unwin, Sydney

Garnham, Nicholas 1990, *Capitalism and communication, global culture and the economics of information*, Sage, London

Hamelink, Cees 1990, 'Information imbalance, core and periphery', *Questioning the media: a critical introduction*, eds John Downing, Ali Mohammadi & Annabelle Sreberny-Mohammadi, Sage, London

Independent Commission Against Corruption 1992, *Inquiry into the sale of government information*, NSW Government, Sydney

Keane, John 1991, *The media and democracy*, Polity Press, Cambridge

Lee, Michael 1992, *News and fair facts, report of the select parliamentary inquiry into the print media*, Australian Government Publishing Service, Canberra

Litchenberg, Judith 1991, *Democracy and the mass media*, Cambridge University Press, Cambridge

Macpherson, C.B. 1977, *The life and times of liberal democracy*, Oxford University Press, London

Melody, William 1990, 'The information in I.T. Where lies the public interest?', *Intermedia*, vol. 18, no. 3, June–July, pp. 10–18

Michaels, Eric 1986, *The Aboriginal invention of television, Central Australia 1982–86*, Australian Institute of Aboriginal Affairs, Canberra

Postman, Neil 1984, *Amusing ourselves to death*, Penguin, New York

Schultz, Julianne 1990, 'Media flowering under watchful eyes: East Europe [and Western media empires]', *Australian society*, October, pp. 8–9.

——1992a, 'Encouraging competition and diversity without offending the monopolists', *Media information Australia*, no. 65, August

——1992b, 'Investigative reporting tests journalistic independence', *Australian journalism review*, vol. 14, no. 2, July–December

Shawcross, William 1992, *Rupert Murdoch*, Chatto & Windus, London

Annotated bibliography

Garnham, Nicholas 1990, *Capitalism and communication, global culture and the economics of information*, Sage, London
A perceptive account of the changing debates about the relationship between

technology, mass media and society. The essays draw on and develop the Habermasian notion of the public sphere and civil society.

Keane, John 1991, *The media and democracy*, Polity Press, Cambridge
Exploring the development of notions of the freedom of the press, within a discussion of the nature of democracy, this collection argues for public service media as an essential component of democratic communications.

Macpherson, C.B. 1977, *The life and times of liberal democracy*, Oxford University Press, London
Macpherson locates the development of liberal democracy within an economic framework, and is optimistic about the democratic possibilities of participatory technologies.

Schultz, Julianne 1992a, 'Encouraging competition and diversity without offending the monopolists', *Media information Australia*, no. 65, August
My analysis of the politics behind the only national inquiry into the print media looks at the realpolitik of relations between the media corporations and the politicians and the limited possibilities for regulation, change and intervention. It critically assesses the shortcomings of the inquiry and its recommendations.

Shawcross, William 1992, *Rupert Murdoch*, Chatto & Windus, London
Shawcross's book documents the development of News Ltd and identifies Murdoch's early realisation of the potential of the information age—together with the way he strategically positioned his company to be one of the big players in the global information industry.

DATAVEILLANCE: DELIVERING *1984*
Roger Clarke

During the last few decades, information technology (IT) has become highly sophisticated, and real benefits have been achieved. There have also been considerable negative impacts.

This chapter discusses the application of IT to the surveillance of people through their data. It argues that the risks to individuals and society as a whole are enormous. Moreover, intrinsic control mechanisms are entirely inadequate to ensure measured and balanced application of the techniques of surveillance, and extrinsic control measures have been inadequate and too late. The power relationship between large data-dependent organisations and members of the public is now so unbalanced that concerted action is necessary if the important value of information privacy is to be sustained into the twenty-first century.

1984 is arriving a little late

During the last 70 years, a series of anti-utopian novels has chilled us with descriptions of imagined futures in which governments use information technology to exercise control over society. Eugene Zamyatin's *We* established the genre as early as 1922. An important recent example is John Brunner's *The shockwave rider*. But the image that forms our nightmares is George Orwell's *1984*. In the early 1980s it was fashionable to pretend that the year had been a prediction (it was not: literary critics think that it was merely a reversal of the digits in the year in which Orwell finished the novel—1948). By 1984 some serious incursions into privacy had occurred, and had not been corrected; but the advanced western societies that are Australia's reference point were still far freer than the constrained world that the tormented Winston Smith endured. In the mid-1990s they still are. But, as a few examples will show, the rein has been tightening quickly during the last decade.

From the late 1950s onwards, corporations and government agencies have made increasing use of information technology. For twenty years computers were mainly used to address needs within individual organisations. Gradually, however, the capacity of machines grew to the point where it was feasible to share data between systems in different organisations, and even to conceive systems which crossed organisational boundaries.

Many such systems involved data about human beings. It proved difficult to combine data from different systems, because each system identified the people it dealt with using different schemes; for example, each finance company had its own customer code, and each government agency its own client or file number. As Representative Frank Horton commented, in a submission to the 1966–7 hearings on a proposed national data centre in the US:

> One of the most practical of our present safeguards of privacy is the fragmented nature of present information. It is scattered in little bits and pieces across the geography and years of our life. Retrieval is impractical and often impossible. A central data bank removes completely this safeguard (Frank Horton, cited in Rule et al. 1980, p. 56).

Corporations are driven by the profit motive, and hence efficiency in the usage of resources is important to them. During the Labor government of the 1980s, efficiency also became an important motivator in the Commonwealth public sector. In March 1985 the suggestion emerged that all manner of evils could be combated if a single identifier were used for dealings with a large number of organisations. The proposal was to create a register (which the government tried to argue was not a centralised database) whereby participating agencies could share specific data about individuals. The entire population was to be recorded on the register and the national identification scheme was designed to harness patriotic fervour, by calling it the Australia Card and decking it out in green and gold (Clarke 1992a, p. 37–9).

For two-and-a-half years, the government battled to have its proposal embraced by an apathetic public, and accepted by a Senate in which it was in the minority. They claimed that huge savings would be made from reeling in tax avoiders, social security cheats and illegal immigrants, and that anyone who opposed the scheme must have something to hide. (See Greenberg 1984; Greenberg et al. 1986; Kusserow 1984; SMOS 1987 and JSCAC 1986 for arguments in favour of government agencies sharing data on individuals.)

Critics ridiculed the bureaucrats' estimates. They also drew attention to the vast amount of private information which would become available to thousands of public servants, and the significant change in the balance of power between individuals and the state which the scheme would bring about (Clarke 1987).

At first the Opposition and the Democrats (who held the balance of power) were sceptical, but as more information became available and more analysis was undertaken, they became virulently opposed to the proposal.

Finally the import of the scheme became apparent to the public, and after street marches and letters to the editors of newspapers throughout the country, the Australia Card scheme was withdrawn.

Within months, the government brought forward an alternative proposal involving enhancements to the Tax File Number (TFN), to enable the objective of reducing tax avoidance to be addressed. This enactment was accompanied by Australia's first privacy statute, and passed into law at the end of 1988. The TFN proposal had not been subjected to rigorous cost–benefit analysis. As Laudon notes, 'a pattern has emerged among executive agencies in which the identification of a social problem [such as tax evasion] provides a blanket rationale for a new system without serious consideration of the need for a system' (Laudon 1986, p. 385).

The TFN enhancements, which allowed computer matching of data, were approved by parliament on the basis that only the Australian Taxation Office would ever use the number. Within a year, a significant number of additional uses had surfaced, dummied through an inattentive parliament in various ways. During 1990, the government proposed extension of TFN use to the entire welfare sector. The recession had deepened, and the Opposition had moved decisively toward the right. Social factors now paled into insignificance against the dominant economic concerns, an accommodation was reached between the parties, and the legislation passed. Further, with both sides of parliament intent on supporting social control measures, public service executives grasped their opportunity and proposed a very large-scale computer matching program. This was duly enacted subject to substantial (if flawed) regulatory measures.

The regulatory measures raise an important issue: does privacy legislation—or the existence of a watchdog—serve to facilitate the spread of surveillance by making it appear less dangerous? The test of data protection legislation (and privacy watchdogs) is not the extent to which they regulate established mechanisms of data surveillance so as to make them operate less invasively—or more 'efficiently'—but rather the capacity that they have to prevent unacceptably dangerous systems being established in the first place (Flaherty 1989).

A variety of other programs have been implemented by the Commonwealth public sector during the last few years which have involved significant intrusiveness into people's affairs. In 1990, an obligation was placed on financial institutions to report all cash transactions above the value of $10 000. As was the case with the 'tax' file number, 'function creep' occurred, and the 'cash transactions' reporting scheme was soon extended to apply to a variety of non-cash transactions.

During the period 1990–93, the Attorney-General's Department developed proposals for the Law Enforcement Access Network (LEAN), a scheme with data analysis capabilities powerful enough to support the work of professional investigators, but available to in excess of 10 000 public servants. For many months the department claimed it was not subject to the Privacy Act, and when the department relented on this point it claimed instead that it was subject to that Act but covered by a series of exemptions which had the effect of rendering the privacy watchdog powerless (Clarke 1992b). In 1992, another proposal with potentially huge privacy ramifications emerged, relating to a Health Communications Network.

Two further examples of government programs highlight a couple of important points. An exhaustive investigation by the NSW Independent Commission Against Corruption (ICAC) in the period 1990–92 identified many instances in which individuals sold nominally confidential personal data for their own advantage. Much worse, however, was the finding that many banks, other companies and government agencies had been active participants in trade which was in all cases highly morally dubious, and in many cases technically illegal (ICAC 1992).

Meanwhile, on several occasions during the late 1980s and early 1990s, the Health Insurance Commission (which, among several other things, operates the Medicare program) brought forward a proposal to gather centrally in Canberra all details of all prescriptions issued in Australia. Its purpose was to address fraud and over-servicing by medical practitioners and excessive discounts on prescription items which it claimed were being granted to patients at taxpayer expense. Privacy advocates argued that the very high degree of privacy-invasiveness inherent in the proposal was virtually ignored, and that the estimates of net financial benefits arising were outlandish. In one of the rare instances of a proposal being subjected to prior review, the Auditor-General and the Department of Finance found the claims over-stated and recommended against the project proceeding. The proposal was withdrawn.

Agencies have shown a great deal of persistence with schemes they judge to be in their own interests, however, and it must be expected that the proposal will re-emerge at a time the public service executives concerned consider auspicious.

Bold and insensitive applications of IT are invading privacy on a scale never before possible. The purpose of this chapter is to look behind the schemes to understand their nature and the context in which they are brought forward. To do so, it is first necessary to review developments in IT, and then appreciate why it is so attractive to organisations in general and government agencies in particular. A brief description is then provided of the various techniques which make up modern 'dataveillance'. Some of the key risks it entails are identified, and actual and potential control mechanisms are discussed, in order to ensure that organisations' practices do not sacrifice humanity in the search for resource efficiency.

The promise of information technology

Information technology (IT) refers to the combination of computing, communications and robotics. Computers were developed in the 1940s, applied to business and government progressively since the 1950s, and married to both local communications and tele- (distance) communications progressively since the 1960s. Robotics, the combination of computers with machines which directly sense and affect their environment, has developed since the 1970s. Information technology, including robotics, can be used to develop systems which cross-match and automatically access data held in a variety of media; including magnetic discettes and tapes, and optical storage such as CD-ROMs. There are many positive uses of IT, and the author of this chapter is actively involved in its development, its application, and the education of further generations of professionals to assist organisations of all kinds to apply it in a profitable and responsible manner.

During the last four decades there has been dramatic growth in the capabilities and capacity of information technology. Processor speed and storage capacity have grown exponentially, and costs have fallen just as steeply. Software development techniques, which for the first two decades slowed down and lowered the quality of applications, have been improving.

It would be wrong to let these explosions in computing blind us to the developments which preceded and ran parallel with the developments in computing, including telex and telefax transmission, and offset printing, photocopying and laser-writing. Since then there have been great improvements in data, voice and now image communications, both locally and over distance. And although developments in robotics have been much slower, there have been real developments also in the integration of intelligence into machines, and sensors and effectors into computing systems.

The need for information technology

Even after the early waves of the industrial revolution, the scale of economic activity was low by modern standards, and the governmental and private sector institutions which undertook such activities tended to be small. With the dramatic increase in scale of economic activities during the present century has come a dramatic increase in the 'social distance' separating individuals from the institutions with whom they transact the majority of their business. This social distance can be thought of as the level of distrust felt by both parties, such as between individuals and bureaucrats who deal with large numbers of the great unwashed public.

To make up for the loss of immediacy in dealings between people who knew one another, there has been a great increase in the data-intensity with which organisations operate—government agencies and companies alike now depend very little on the judgment of employees local to the individual concerned and very heavily on the information that they store in their files and use centrally to the organisation but remotely to the individual. Information privacy is a relatively recent preoccupation, and in the mid-1980s a senior Australian Cabinet Minister went so far as to denigrate it as a 'bourgeois value'. Until the last few decades, it was not necessary for people to express concern about it, or for parliaments to create laws protecting it. This was because of the highly dispersed nature of data storage, the difficulty of finding data when it was wanted, and the difficulty of copying and transmitting the data once it was found; in other words, information privacy was protected by the enormous inefficiency of data handling.

IT has progressively reduced that inefficiency, to the great benefit of organisations and their clients. Concomitant with those efficiency improvements, however, has been the disappearance of the traditional information privacy protections which relied upon data being dispersed and difficult to collect and analyse. New mechanisms are needed to ensure that society's headlong rush for efficiency does not mortally wound other critically important human values.

Dataveillance

This section discusses a class of applications of information technology referred to as dataveillance. By this is meant automated monitoring through computer-readable data rather than through physical observation. Although the techniques are as applicable to goods proceeding along a production line as to people, this chapter restricts its attention to the surveillance of humans. Dataveillance is of real potential benefit; for example in the detection of individuals who are worthy of attention, possibly because they are in need, or because they represent a threat to others. The focus of the paper is, however, largely on dataveillance's 'downside'. (See e.g. Rule 1974; Rule et al. 1980; Burnham 1983 and Roszak 1986 for general reviews of data surveillance, and Kling 1978 for one of the foundation works in the field.)

Surveillance is the systematic investigation or monitoring of the actions or communications of one or more persons. Rule uses the term for 'any form of systematic attention to whether rules are obeyed, to who obeys and who does not, and how those who deviate can be located and sanctioned' (Rule 1974, p. 40). Surveillance has traditionally been undertaken by physical means such as prison guards on towers. In recent decades it has been enhanced through image-amplification devices such as binoculars and high-resolution satellite cameras. Electronic devices have been developed to augment physical surveillance and offer new possibilities such as telephone 'bugging'.

The last 25 years have seen the emergence and refinement of a new form of surveillance, no longer of the real person, but of the person's data-shadow, or digital persona. Dataveillance is the systematic use of personal data systems in the investigation or monitoring of the actions or communications of one or more persons. It may be 'personal dataveillance', where a particular person has been previously identified as being of interest. Alternatively it may be 'mass dataveillance', where a group or large population is monitored, in order to detect individuals of interest, and/or to deter people from stepping out of line.

A variety of techniques exists (see Exhibit 9.1). Front-end verification (FEV), for example, comprises the checking of data supplied by an applicant (e.g. for a loan or a government benefit) against data from a variety of additional sources, in order to identify discrepancies. FEV may be applied as a personal dataveillance tool where reasonable grounds exist for suspecting that the information the person has provided may be unreliable; where, on the other hand, it is applied to every applicant, mass dataveillance is being undertaken. Data matching is a facilitative mechanism of particular value in mass dataveillance. It involves trawling through large volumes of data collected for different purposes, searching for discrepancies and drawing inferences from them.

Exhibit 9.1 Dataveillance techniques

Personal dataveillance, of previously identified individuals

- integration of data hitherto stored in various locations within a single organisation
- screening or authentication of transactions against internal norms
- front-end verification of transactions that appear to be exceptional, against data relevant to the matter at hand, and sought from other internal databases or from third parties
- front-end audit of individuals who appear to be exceptional, against data related to other matters, and sought from other internal databases or from third parties
- cross-system enforcement against individuals, where a third party reports that the individual has committed a transgression in his or her relationship with the third party

Mass dataveillance, of groups of people

- screening or authentication of all transactions, whether or not they appear to be exceptional, against internal norms
- front-end verification of all transactions, whether or not they appear to be exceptional, against data relevant to the matter at hand, and sought from other internal databases or from third parties
- front-end audit of individuals, whether or not they appear to be exceptional, against data related to other matters, and sought from other internal databases or from third parties
- single-factor file analysis of all data held or able to be acquired, whether or not they appear to be exceptional, involving transaction data compared variously against a norm, permanent data or other transaction data
- profiling, or multi-factor file analysis of all data held or able to be acquired, whether or not they appear to be exceptional, variously involving singular profiling of data held at a point in time, or aggregative profiling of transaction trails over time

Facilitative mechanisms

- computer or data matching, in which personal data records relating to many people are compared in order to identify cases of interest
- data concentration, the combination of personal data through organisational merger or by the operation of data-interchange networks and hub systems

Risks inherent in data surveillance

Data surveillance's broader social impacts can be categorised as in Exhibit 9.2. By way of example, individuals can suffer as a result of misunderstandings about the meaning of data on the file, or because the file contains erroneous data which the individual does not understand, and against which he or she has little chance of arguing without finding and hiring a specialised lawyer. Such seemingly small, but potentially very frustrating and infuriating personal problems can escalate into widespread distrust by people of government agencies, and of the legal system as a whole. The lack of attention paid to impacts other than apparent improvements in efficiency results in an unacknowledged social cost of dataveillance. Cost–benefit analyses rarely include non-financial cost and benefits, or even description of the non-quantifiable factors.

Exhibit 9.2: Risks inherent in dataveillance

In personal dataveillance

- low data quality decisions
- lack of subject knowledge of, and consent to, data flows
- blacklisting
- denial of redemption

In mass dataveillance
- risks to the individual
  - arbitrariness
  - acontextual data merger
  - complexity and incomprehensibility of data
  - witch hunts
  - *ex–ante* discrimination and guilt prediction
  - selective advertising
  - inversion of the onus of proof
  - covert operations
  - unknown accusations and accusers
  - denial of due process

- risks to society
  - prevailing climate of suspicion
  - adversarial relationships
  - focus of law enforcement on easily detectable and provable offences
  - inequitable application of the law
  - decreased respect for the law and law enforcers
  - reduction in the meaningfulness of individual actions
  - reduction in self-reliance and self-determination
  - stultification of originality
  - increased tendency to opt out of the official level of society
  - weakening of society's moral fibre and cohesion
  - destabilisation of the strategic balance of power
  - repressive potential for a totalitarian government

Clearly, many of these concerns are diffuse. On the other hand, there is a critical economic difference between conventional forms of surveillance and dataveillance. Physical surveillance is expensive because it requires the application of considerable resources. With a few exceptions (such as East Germany under the *Stasi*, Romania, and China during its more extreme phases), this expense has been sufficient to restrict the use of surveillance. Admittedly the selection criteria used by surveillance agencies have not always accorded with what the citizenry might have preferred, but at least its extent was limited. The effect was that in most countries the abuses affected particular individuals who had attracted the attention of the state, but were not so pervasive that artistic and political freedoms were widely constrained.

Dataveillance changes all that. Dataveillance is relatively very cheap, and getting cheaper all the time, thanks to progress in information technology. The economic limitations are overcome, and the digital persona can be monitored with thoroughness and frequency, and surveillance extended to whole populations. To date, a number of particular populations have attracted the bulk of the attention, because the state already possessed substantial data-holdings about them. These are social welfare recipients and government employees. Now that the techniques have been refined, they are being pressed into more general usage, in the private as well as the public sector.

Controls

If dataveillance is burgeoning, controls are needed to ensure that its use is not excessive or unfair. There is a variety of natural or intrinsic controls, such as self-restraint and morality. Unfortunately morality has been shown many times to be an entirely inadequate influence over people's behaviour (Clarke 1992c; and e.g. the quarterly newsletter of the US-based Computer Professionals for Social Responsibility—CPSR). There is also the economic constraint, whereby work that is not worth doing tends not to get done, because people perceive better things to do with the same scarce resources. Regrettably this too is largely ineffective. Cost–benefit analysis of dataveillance measures is seldom performed, and when it has been, it has been restricted to factors which can be reduced to fundamental measures. Further, the quality of such cost–benefit analyses has generally been appalling. 'Project Match' was an early American computer matching program carried out in 1977:

> Project Match compared 'the records of roughly 78% of all recipients of Aid to Families with Dependent Children (AFDC) with the payroll records of about 3 million federal employees'. It identified 33 000 raw hits, later reduced to

7100, resulting in 638 internally investigated cases, of which 55 resulted in prosecutions (OTA 1986, p.42) . . . Project Match was claimed to be a great success. It appears, however, that these prosecutions resulted in only about 35 convictions, all for minor offences, with no custodial sentences and less than $10 000 in fines. This paradox of a project being hailed as a great success when the measurable financial costs are high and the measurable financial benefits very low, has been a feature of matching programs from the very beginning (Clarke 1992d, p. 8).

This example reflects the dominance of political over economic considerations—both politicians and public servants want action to be seen to be being taken, and are less concerned about its effectiveness than its visibility.

If intrinsic controls are inadequate, extrinsic measures are vital. For example, the codes of ethics of professional bodies and industry associations could be of assistance (Clarke 1990). Regrettably, these are generally years behind the problems, and largely statements of aspiration rather than operational guidelines and actionable definitions of what is and is not acceptable behaviour. Over twenty years after the information privacy movement gathered steam, there are few and very limited laws which make dataveillance activities illegal, or which enable regulatory agencies or the public to sue transgressing organisations. A (limited) statute exists at national level in Australia, but none at all at the level of State governments. In any case, statutory regimes are often weak due to the power of data-using lobbies, the lack of organisation of the public, and the lack of comprehension and interest by politicians. The public has demonstrated itself as being unable to focus on complex issues; public apathy is only overcome when a proposal is presented simply and starkly, such as 'the state is proposing to issue you with a plastic card. You will need to produce it whenever anyone asks you to demonstrate that you have permission to breathe'.

It is clear that the dictates of administrative efficiency are at odds with individual freedoms, and that the power of dataveillance techniques is far greater than it was a decade or more ago. It is essential that governments consider each dataveillance technique and decide whether it should be permitted under any circumstances at all; if so, what those circumstances are, what the safeguards should be and what control mechanisms will ensure that each of these safeguards operates effectively and efficiently (Clarke 1991, pp. 514–19).

There is a tendency for dataveillance tools to be developed in advanced nations which have democratic traditions and processes (however imperfect). There is a further tendency for the technology to be exported to less developed countries. Many of these have less well developed democratic traditions, and more authoritarian and even repressive regimes. The control mechanisms in advanced western democracies are inadequate to cope with sophisticated dataveillance technologies; in third world countries there is very little chance indeed of new extrinsic controls being established to ensure balance in their application. It appears that some third world countries may be being used as test-beds for new dataveillance technologies.

Conclusion

There continues to be much that information technology can offer to improve people's lives. The success of IT has, however, created a very serious risk that the drive for efficiency will seriously damage human values. The enthusiasm of our institutions for IT must be tempered if we are to avoid that damage. Intrinsic protections have proven inadequate, and the country's parliaments have shown themselves time and again to be stadia for bull-fighting, not for consideration of laws and economic and social priorities. In the absence of concerted action by individuals and public interest advocacy groups, *1984* will arrive; just a little late.

References

Brunner, John 1988, *The shockwave rider*, Methuen, London

Burnham, D. 1983, *The rise of the computer state*, RandomHouse/Weidenfeld & Nicholson, New York

Clarke, Roger A. 1987, 'Just another piece of plastic for your wallet: the "Australia Card" scheme', *Prometheus*, vol. 5, no. 1, June, pp. 29–45, republished with an addendum 1988, *Computers and society*, vol. 18, no. 3, July

——1990, 'Social implications of IT—the professional's role', *Australian computer journal*, vol. 22, no. 2, May, pp. 27–9

——1991, 'Information technology and dataveillance', *Computerisation and controversy*, eds C. Dunlop & R. Kling, Academic Press, New York, pp. 496–522 (previously published in 1989, *Communications of the ACM*, Association for Computing Machinery Inc., vol. 31, no. 5, May, pp. 498–512 and republished in 1993, *Information technology and social issues*, eds C. Huff & T. Finholt, McGraw-Hill, New York

——1992a, 'The resistible rise of the national personal data system', *Software law journal*, vol. V, no. 1, February, pp. 29–59

——1992b, 'LEAN times ahead?', *Policy*, vol. 8, no. 2, Winter, pp. 55–7

——1992c, 'Privacy needs more than good intentions: the lessons of the ICAC enquiry', *Professional computing*, November, pp. 4–10

——1992d, *Computer matching by government agencies: a normative regulatory framework*, working paper, Department of Commerce, Australian National University, Canberra

Flaherty, D. 1989, *Protecting privacy in surveillance societies*, University of North Carolina Press, Chapel Hill

Greenberg, D.H. 1984, 'Sniffing out fraud: computer sleuthing comes to public welfare', *Public welfare*, Summer, pp. 32–9

Greenberg, D.H. and Wolf, D.A. (with Pfiester, J.) 1986, *Using computers to combat welfare fraud: the operation and effectiveness of wage matching*, Greenwood Press Inc., US

ICAC 1992, *Report on unauthorised release of government information*, 3 vols, Independent Commission Against Corruption, Sydney, August

JSCAC 1986, *Report of the Joint Select Committee on an Australia Card*, 2 vols, Australian Government Publishing Service, Canberra, May

Kusserow, R. P. 1984, 'The government needs computer matching to root out waste and fraud', *Communications of the ACM*, Association for Computing Machinery Inc., vol. 27, no. 6, June, pp. 542–5

Kling, R. 1978, 'Automated welfare client tracking and welfare service integration: the political economy of computing', *Communications of the ACM*, Association for Computing Machinery Inc., vol. 21, no. 6, June, pp. 484–93

Laudon, Ken C. 1986, *Dossier society: value choices in the design of national information systems*, Columbia University Press, New York

Orwell, George 1972, *1984*, Penguin Books, New York (originally published in 1948)

OTA 1986, *Federal government information technology: electronic record systems and individual privacy*, OTA–CIT–296, US Congress, Office of Technology Assessment, US Government Printing Office, Washington DC, June

Roszak, Theodore 1986, *The cult of information*, Constable, London

Rule, J.B. 1974, *Private lives and public surveillance: social control in the computer age*, Schocken Books, New York

Rule, J.B., McAdam, D., Stearns, L. and Uglow, D. 1980, *The politics of privacy*, New American Library, New York

SMOS 1987, *Review of systems for dealing with fraud on the Commonwealth*, Department of the Special Minister of State, Australian Government Publishing Service, Canberra, March

Zamyatin, Eugene 1983, *We*, Penguin Books, New York (originally published Russ. 1920; Eng. trans. 1922)

Annotated bibliography

Clarke Roger A. 1992d, *Computer matching by government agencies: a normative regulatory framework*, working paper, Department of Commerce, ANU, Canberra
Reporting upon the development and application of techniques in Australia

and the United States, the paper proposes a framework for regulation which includes general principles and detailed requirements.

Davies, S. 1992, *Big brother: Australia's growing web of surveillance*, Simon & Schuster, Sydney
Combining history, interpretation, analysis and prediction, Davies is gloomy about the prospects for Australia unless we sustain a belief that technology is meant to serve humans and not the other way around.

Flaherty, D. 1989, *Protecting privacy in surveillance societies*, University of North Carolina Press, Chapel Hill
A careful assessment of the performance of parliaments in several countries, concluding that no country has done well, and that the United States has done very badly.

Laudon, Ken C. 1986, *Dossier society: value choices in the design of national information systems*, Columbia University Press, New York
A key work on dataveillance and privacy.

NSWPC, 1976–94, *Annual reports*, and reports on particular issues as available, NSW Privacy Committee, Sydney
Australia's oldest permanent 'watchdog' agency and the only one at State level. Its complaint-handling and research have been of high quality.

Privacy Commissioner 1989–94, *Annual reports*, and reports on particular issues as available, Human Rights & Equal Opportunities Commission, Sydney
The recently created 'watchdog' agency at national level. Better funded than the NSW Committee, it has produced high-quality research, but is constrained by legislation, and appears less inclined to champion the privacy interest in the face of Government policy measures.

ELECTRONIC NEIGHBOURHOODS: COMMUNICATING POWER IN COMPUTER-BASED NETWORKS

Lynda Davies and Wayne Harvey

Computer-based networks (CBNs) in this information age are so prevalent and pervasive that they are virtually invisible to all except those who implement them. Used as an alternative to other slower, manual methods, they are essentially a system for the direct transfer of data between separate computers. The technology which allows this interconnection to occur is simple in concept whilst complex in implementation. It is becoming so widespread that it is feasible to consider that within not very many years almost every computer on the planet will be interconnected in a global network. This chapter describes CBN—the network as an entity; network politics and power; the problems and effects of this globalisation; and a short case study which highlights the political nature of a form of CBN.

Network technology

One may think of a network as simply a cable that connects one or more computers, such that they may transfer data to one another. In essence, this is true, but it is not that simple. Along with the cabling, which must adhere to certain standards (pin connections, shielding etc.), come the protocols for that transfer—what the format of the message to be transferred must be in order for it to be sensibly interpreted by the receiving computer.

There are many protocols in existence, with most being incompatible with any other, but globally there are some *de facto* standards developing: for example token ring topologies for local area networks, TCP/IP and ethernet for wide area networks. Furthermore, the software which prepares the data for transfer must also adhere to standards relative to the network hardware in use, otherwise the data transfer will become garbled before transmission. Lastly, there is the data which is to be sent. If one computer prepares a sales order request and sends it to a banking computer that expects accounts numbers and a transaction, it will not make sense. Hence, the network must be sensibly interconnected, or the message must include a destination address encoded within it.

Network usage

The information being transferred in CBN conforms to two main types.

Machine-readable information

Machine-readable information is transferred directly between two or more machines without humans interacting with the information during the process. One machine formats and transmits a request from its user or software, which is received by a second machine, interpreted, acted upon and stored for future retrieval, or returned to the originating computer. An example is a sales order which, after confirmation, is returned to the buyer's computer, along with data about the product requested. The sale is then concluded along another CBN channel—the seller's computer requests a banking computer to debit the buyer's account by the value of the sale, and both buyer and seller are notified of the success of the transaction. Such multiple transactions to perform one function are commonplace with CBN, and the speed at which they are performed means that the sale can be concluded almost instantaneously.

Human-readable information

Human-readable information such as E-mail, fax services, file transfer, news services, direct conversations and similar, are transferred directly from sender to recipient. This form of CBN allows instantaneous communication between users of the network on a personal level, regardless of the location of either party. This information is assembled by the sender, interpreted by the sender's machine into a form that is compatible with the network, sent, and interpreted by the receiver's machine into a human-readable form for presentation. The globalisation of this service means that personal communication need not be within the same time zone. The recipients need not be at their machines to receive, except in the case of direct conversations, where both parties are conducting a live 'conversation' via the electronic media.

In this chapter we concentrate on the use of human-readable information in socio-political contexts. An important, related issue arising from increased organisational dependency on CBN is data transfer overload. The quantity of data that is being transferred across networks is continually on the increase, with some of the rapidly emerging factors being:

- the number of computers connected;
- the variety and quantity of information being transferred between these computers;
- private messages between individuals, as described above; and

- the network being used for pleasure and recreation purposes by individuals.

These factors are leading towards a global network in which information overload (where the quantity of information available to an individual or organisation becomes too great for sense to be made of it) becomes a peril. Thus, the users of the network must put in place 'filters' for the information, which remove from their access information which is of no consequence to them (Judkins, West & Drew 1985). Mostly, this function is performed by the systems operators using the hardware or software of the network, by barring access to certain computers, creating private networks, establishing password codes to private information and restricting the information that is stored on machines connected to the network.

Modelling a communications network

Computer-based networking can be viewed from a technical perspective or from the perspective of its impact on the human context (Forester 1989; Kiesler & Sproull 1987; Sproull & Kiesler 1991). We take a radical view from the human perspective and state that CBN can only happen as part of a communications network. This means that a message has to be transferred to a receiver in a way which makes the end result both sensible and meaningful. The important part of any CBN process is sense-making and the construction of meaning. CBN is a particular kind of human communication, particular in that it relies upon computer-based data transfer. CBN is also a special case of general human communication, and the relevant question is really concerned with the way that CBN is special.

Current models of CBN are aimed at two audiences, the technologists and the managers. At present there is no general theory seeking to explain CBN such that any interested audience could take part in the discussion. We have found it necessary to look at CBN in a more philosophical manner so that we can understand the principles of any communications network before looking at CBN in particular. Our four-dimensional interpretation of networking is intended to avoid technical and managerial jargon whilst exploring some first principles of communication networking. We seek to develop knowledge of networking that is freed from the domination of current power groups through our use of a more philosophical perspective to look at CBN.

CBN is dependent upon shared understanding and common knowledge, specialised because of its reliance upon information technology. CBN is driven by human intentions, with those intentions underpinning the use of technology to control communications and gain power. By analysing the

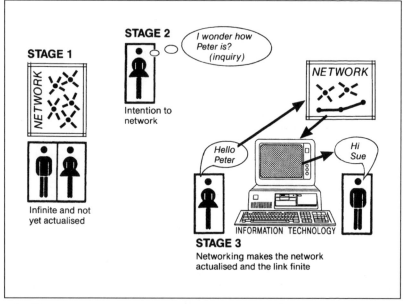

FIGURE 10.1 Conceptualising a network

political use of CBN, we make power and CBN more visible and hence more socially accountable.

Imagine an entity which is not tied to physical time and space. According to some philosophers, the human mind is such a form. This space has abilities to structure itself but is only known when it is made 'real', that is, when someone or some group, makes it become an actual part of reality (actualises it). All around us, at all times, this four-dimensional space is capable of becoming a network in three-dimensional reality. However, there are constraints on how this can happen and these constraints mean that any use is finite.

To use this space, someone must try to contact someone else—in other words, there has to be an intention. When the initiator contacts the intended person, a network is formed; and a network formation exists as long as both parties can make sense of the message being conveyed. Often information technology is used to aid this process. A simple illustration of the process is shown in Figure 10.1.

To make sense of messages being conveyed, a number of different things happen at once. The intention has to be sensibly interpreted so that there is a common understanding; the content must be formed (encoded) and interpreted (decoded). For this to happen, understandings of how to interpret messages have already to be present. This shows that there has to be shared knowledge existing between people for networking to be possible. Potentially, knowledge is everywhere. In CBN, knowledge potentials are represented as *nodes*, and *relationships* are formed between these nodes when some intention to network is acted out. For the network to form, the nodes and their relationships have to share a common interpretation of how the knowledge clusters created are meaningful. We have called the structure which links together particular knowledge nodes in a sensible manner a *paradigm*.

The process of turning the potential of a CBN into a networked reality is restricted by the need for the communicators to share a paradigm. Networked communication can only sensibly occur through the guidance of paradigms. For example, when Sue contacts Peter (see Figure 10.1), she assumes that Peter will have knowledge of how to receive her contact and will have some expectation of what her communication intentions are. This demands a common understanding of the language being used and reasonable knowledge of how to use the information technology supporting the network transaction. This variety of knowledge restricts who can be part of a particular network because those without a minimum working knowledge of all of the above cannot understand the process and so are outside that particular knowledge paradigm.

Computer-based information exchange networks realise their potential in different ways. In principle the CBN is infinite, anyone could be a user of a network. However, if individuals do not have adequate understanding (a minimum working knowledge) of how to send or receive a contact, which includes access to the appropriate technology and knowledge about how to use that technology effectively, they will be excluded from entry to communication in that particular paradigm.

Lack of knowledge of the language can be a major excluder in computer-based networking. For instance, the user may need to know that a Unix operating system is at the core of the process in order to select the right terms (language) to communicate with the computer. Knowing what part of the software system is in operation at any part of a transaction is often also necessary. A poor 'user interface' (the communication structure designed as a linkage between the world of the computer, and the world of the humans using the computer) can prevent knowledge from being available during the network transaction, thus excluding some people from that knowledge transaction.

Many CBN systems have their technical knowledge transactions well hidden (invisible) so that only those people with in-depth technical knowledge can interact in that network. However, this is not a necessary feature of CBNs, many are more visible and more accessible. The invisibility can be manipulated to form paradigm barriers to entry and use of a network. Power to create invisibility is effective in access control.

The human context of CBN

Any network happens because humans desire to interact with other humans. Even machine-readable transfer of information only occurs because there has been a human need for that transference, and it is people who design that need into a computer-supported form. Networks are communication operations which occur because of human intentions, needs and desires (Rosnow 1992).

Looking at networks, therefore, demands analysis of the human intentions which support, enhance, or steer the development of CBN systems. Because of the knowledge constraints associated with CBN networks, and discussed in the previous section, there is the potential for power manipulation in the use of any CBN system. Those who have the knowledge can design into the CBN exclusion of those who do not have the knowledge. CBN becomes a tool for the creation and manipulation of knowledge-objects as a means of controlling, disempowering and manipulating others (Johansen and DeGrasse 1992).

Many sociologists have sought to explain the effects of new technology on political processes. One in particular, Michel Foucault (1974), argues that knowledge is used as a political tool, through the creation and manipulation of knowledge-objects which lead to the exclusion of the non-knowledgeable from communication exchanges. Foucault explains that knowledge creates power which, through political actions, circumscribes those who have access to the knowledge. The forms of knowledge created by this process are named and used in the language of those allowed to share the knowledge, and that naming process creates knowledge-objects. In the hands of the knowledgeable, machines are the means for heightening the politics of knowledge, creating greater restrictions and more opportunities for political manipulation of others who are less knowledgeable (Kiesler and Sproull 1987). In this way, machines are themselves knowledge-objects in the knowledge domain of networks.

Information technology in general, and CBN in particular, can be described as the means for producing and manipulating knowledge-objects. Foucault (1974, 1976, 1977), conceptualises knowledge-objects as creators and products of political domains, and argues that any communication process is dependent upon existing knowledge. Those who create the communication process do so in order to take part in a knowledge use and generation exercise, albeit not always fully intentionally (Tichy, Tushman and Fombrum 1992). The very act of doing this creates a political context where some will benefit due to their ability to interact in a particular knowledge paradigm, and others will be victimised due to their inabilities.

Power politics in CBN: a case study

What is taken to be knowledge is highly variable. Sometimes, those with in-depth technical knowledge can become restricted to technical agendas by their lack of intricate knowledge of the operations of an organisation. Because they are excluded from access to the procedures (committees, informal groups, memos etc.) which give such organisational knowledge, they cannot apply their technical knowledge effectively. They may be purposely excluded because of the depth of their technical knowledge—with this being perceived as a threat by those with more organisational knowledge but with less technical knowledge.

In such an instance the technical knowledge nodes become less significant than the organisational knowledge nodes, because the politically dominant organisational knowledge overrides that of the technical knowledge. Both may be dealing with CBN but the organisational knowledge-object is the more powerful in this example of a knowledge-politics clash.

The study is of an event in an Australian university campus in which a students' association was set up by a small group of students to be a spokesgroup for student computing needs. The students who initiated and ran the association were technically highly competent in the area of computer systems, particularly distributed wide-area computer networks (those which are dispersed over an area greater than a few hundred metres, up to full global networks). One student in particular was highly enthusiastic, with a great deal of self-taught knowledge. He may be described as a gifted amateur who is something of a hacker.

The student association sought support from the information technology (IT) section of the university, foreseeing no rational reasons why such support should not be willingly given. The members of the IT section were paid professionals as opposed to gifted amateurs. The members of the student association did not realise the extent to which they could be seen as a threat by the IT section. The students had real-world technical expertise which was greater than that of the paid IT professionals—the members of the university section with whom they were interacting.

The manager of the IT section continually put additional constraints on the key members of the student association. The representatives from the student association had managed to negotiate limited positions in certain committee meetings held by the IT section. They had been allowed to be present at the general discussion sections of one policy board meeting, but were excluded from the latter sections of that meeting which dealt with the finer details of user policy. (They were requested to leave the meeting prior to this point.) They had also been allowed access to the manager and an operator to discuss purchasing policy, but were not allowed to submit documents recommending purchasing initiatives. They could request information on access to networks but were often misinformed, or given no information, or even implicitly threatened with exclusion from further meetings, if they continued to request information on access restrictions.

The main area of conflict was over students' access to news networks on the E-mail. This was said to be a security issue, yet no explanation was given as to how it could be so. Rather, it seemed that students were defined as a non-voice, made invisible/unheard through the device of renaming student access as a security issue. The proposed access was also said to involve unacceptable expense whilst, in actual fact, E-mail is extremely cheap to run. Further, it was argued that students created an unfeasible administrative load which was not so, as their use simply backed onto the academics' use of the network. Another argument was that access to the news network was a luxury, a diversion from students' learning, despite the fact that a large amount of experiential learning occurs in this way. The continual curiosity-based use of E-mail could enhance students' existing technical knowledge, creating a greater threat to those IT professionals without that degree of technical knowledge.

The organisational knowledge was used as a means of underplaying the technical knowledge through the presentation of the argument that the CBN system of E-mail is a research tool and, therefore, should not be made available to non-researchers, such as undergraduate students. The knowledge paradigm clash had resulted in the network becoming renamed as a knowledge-object owned solely by one group of power-defined communicators (the researchers and their administrative support), through stressing the role of the system as a research-only communicative tool.

This renaming of the news networks served politically to exclude the student group from the technology. Paradoxically, the reason for this was the threat of the students' technical knowledge. This threat created a perceived power issue which led to the use of system access restrictions as a means of disempowering those with the greater technical knowledge.

In this case study, information is power only for those who are already members of existing organisational power networks. On the surface, the students were information-rich because of their extensive knowledge of CBN. By comparison, the IT section professionals were information-poor. In order to reverse the order of things, the IT section used their organisational knowledge to prevent the students from entering organisational power groups. In this way, the students had no voice, and without a voice their technical knowledge became redundant. This made them, in practice, information-poor compared to the IT professionals who had both a strong voice and the power to use that voice to exclude others from being heard. By naming the system as a research system, only academics and their administrative support people had an acknowledged right to speak about the system. The knowledge-object of the network was defined by the more powerful people in that organisational context, to exclude the expression of technical knowledge owned by the students.

This case study demonstrates that network technology is defined as a power element according to how it is perceived as a knowledge-object. Also, the existing political prowess of those interacting within a social context restricts actions to change the knowledge-object. A network can be a prime political tool depending upon who has the knowledge to define it as a particular form of knowledge-object.

Accountability in CBNs

Computer-based networking is often perceived as a threat because it seems to be both pervasive and invisible. The real issue is that those who use CBN to manipulate communication contexts need to be made more accountable. In order to do this, a framework for understanding CBNs as knowledge networks is needed so that the manipulation of networks can be made more visible, leading to an increased potential for enforcing accountability upon the manipulators.

Standards for CBN are developing, not from consensus between all users, but from a few groups/corporations—the power elites—enforcing their own 'standards' upon other users of the network. As there are many such elites, they tend to create standards by consensus within themselves, counterbalancing each other's proposals until a compromise is reached which all users must adhere to.

The encoding of power structures within CBNs, and the use of CBNs for activities such as dataveillance (discussed by Roger Clarke in chapter 9), demonstrate the potential of CBN as a socio-political tool of domination and for disaster creation. The increasing dependence of today's lifestyle on the network creates greater opportunities for sabotage and information terrorism (Hearnden 1989). Hence, the need is great for policy-level, morally bound ethical control of the use of CBN.

The OECD (Organization for Economic Cooperation and Development) has provided guidelines to ensure that democracy can be built into networks (OECD 1980). Using their suggested techniques, such as open access, restrictions on the levels of password usage as constraints, and by ensuring the rights of individuals to know which information is available, the guidelines create a technically feasible democracy in the computer-based use of information. However, if the contexts of any CBN do not put the principles into practice (properly, not just in a cosmetic manner) the guidelines become useless. A truly democratic use of CBN requires policies, directives and an organisational culture which express a truly democratic working environment (Arterton, 1989). In this scenario the issues surrounding the information-rich versus the information-poor can be managed more effectively.

Conclusion

To understand CBN in its human, socio-political context, it is first necessary to understand what is meant by a network. A network is an expansion of the human communicative act, using knowledge structures guided by paradigms, which will often use information technology to support the creation of knowledge-objects as political tools for empowering or disempowering individuals or groups.

CBNs can make visible the political interests of the elites they serve, expressing power in an inequitable and threatening manner. Analyses of the use of knowledge paradigms to enforce domination is needed, showing how electronic media have particular power as knowledge-objects. We need to understand what people know about CBN, how that knowledge is changed in different contexts, and how it is used to prevent alternative knowledge taking priority. Considering the global impact of CBN, this is not a trivial task but it is one of high historical relevance. CBN as a political tool, using knowledge as a political pawn, is a key issue for the social analysis of new technologies.

References

Arterton, F. Christopher 1989, 'Teledemocracy reconsidered', *Computers in the human context*, ed. T. Forester, Basil Blackwell, Oxford, pp. 427–37

Forester, T. 1989, *Computers in the human context*, Basil Blackwell, Oxford

Foucault, Michel 1974, *The archaeology of knowledge*, Tavistock Publications, London

———1976, *The birth of the clinic*, Tavistock Publications, London

———1977, *Discipline and punish: the birth of the prison*, Penguin, London

Hearnden, Keith 1989, 'Computer criminals are human, too', *Computers in the human context*, ed. T. Forester, Basil Blackwell, Oxford, pp. 415–26

Johansen, Robert and DeGrasse, Robert 1992, 'Computer-based teleconferencing: effects on work patterns', *Readings in organisational communications*, ed. Kevin L. Hutchinson, Wm C. Brown, Dubuque, pp. 417–26

Judkins, Phillip, West, David and Drew, John 1985, *Networking in organisations: The Rank Xerox experiment*, Gower, England

Kiesler, Sara, and Sproull, Lee 1987, *Computing and change on campus*, Cambridge University Press, Cambridge
——1987, 'The social process of technological change in organisations', *Computing and change on campus*, eds S. Kiesler & L. Sproull, Cambridge University Press, Cambridge, pp. 28–40
OECD 1980, *Guidelines for the protection of privacy and transborder flows of personal data*, Organisation for Economic Cooperation and Development, Paris
Rosnow, Ralph L. 1992, 'Rumour as communication: a contextualist approach', *Readings in organisational communications*, ed. Kevin L. Hutchinson, Wm C. Brown, Dubuque, pp. 173–86
Sproull, Lee and Kiesler, Sara 1991, *New ways of working in the networked organisation*, The MIT Press, Massachusetts
Tichy, Noel M., Tushman, Michael L. and Fombrun, Charles 1992, 'Social network analysis for organisations', *Readings in organisational communications*, eds Kevin L. Hutchinson, Wm C. Brown, Dubuque, pp. 187–201

Annotated bibliography

Eason, K. 1988, *Information technology and organisational change*, Taylor and Francis, London
An accessible introduction to human aspects of IT, this book offers an account of organisational culture's effect upon the implementation of new technologies.

Forester, T. 1989, *Computers in the human context*, Basil Blackwell, Oxford
A good range of readings on IT and social change, Part Four concentrates upon computers and the future.

Hutchinson, K.L. 1992, *Readings in organisational communications*, Wm C. Brown, Dubuque
Covering topical issues such as teleconferencing, this book also deals with the social nature of networks.

Judkins, Phillip, West, David and Drew, John 1985, *Networking in organisations: the Rank Xerox experiment*, Gower, England
Monitoring organisational change emerging during the implementation of a CBN in Rank Xerox, this research analyses how these issues affected use of the network.

Kiesler, Sara, and Sproull, Lee 1987, *Computing and change on campus*, Cambridge University Press, Cambridge
Concentrating upon organisational implications, this classic text reports

research into the implementation of a CBN across the Carnegie-Mellon University campus.

Sproull, Lee and Kiesler, Sara 1991, *New ways of working in the networked organisation*, The MIT Press, Mass.
This book shows how social networks change and new groups form within the working environment of the organisation as the use of the network emerges.

Walton, R. 1989, *Up and running: integrating information technology and the organization*, Harvard Business School Press, Boston, Mass
Theoretical and practical issues associated with the implementation of information technology into organisations, giving a strategic view of the benefits of integrating networks.

Worksafe Australia 1986, *Repetition strain injury: a report and model code of practice*, Australian Government Publishing Service, Canberra.
Overuse injury is one of a number of health concerns which arise from an increasing reliance upon IT in the workplace, and for leisure. This report was a milestone in the public debate on RSI.

PART III

FRAMING THE GLOBAL

14

THE MULTILOCALS: TRANSNATIONALS AND COMMUNICATIONS TECHNOLOGY
Dick Bryan

Transnational corporations (TNCs) appear to be either loved or hated. It seems, for example, that you either love the McDonald's fast food 'experience' or despise McDonald's for their sugar content, their environmental impact, their 'have a nice day' Americanness, their labour relations and their relentless advertising. But few of us are unaffected by their impact. Probably most TNCs are not as high profile in our daily lives as McDonald's, but they are a conspicuous and increasing presence, not least in the field of technology.

This essay seeks to establish where TNCs fit into economic and social processes, their growth, and in particular, their relationship to new communications technology. It will be argued that TNCs play a critical role in the advancement of international capitalism. They are prominent as both the developers of new technology and as its users. Indeed, the operation of their global networks is often contingent upon the facilities provided by recent advances in telecommunications.

But while TNCs have been targets for much criticism, and even abuse regarding their exercise of power, both economic and political, much of this criticism is misplaced. There are real concerns about some of the things done by TNCs—such as in relation to communications technology—and their ability, individually and collectively, to transform circumstances to their advantage. But the question must be asked whether this is the 'fault' of TNCs, or is attributable to a lack of government regulation, or to the broader international capitalist system of which TNCs are but 'representative'. There exists long-standing debate on these matters. But before determining what TNCs can and cannot do, there is need to clarify what they are.

What are TNCs?

The definition of TNCs is by no means clear—even the United Nations Centre on Transnational Corporations cannot settle on a definition (United Nations 1978). In some sense, it is a matter of knowing one when you see it—as with McDonald's; but this serves to give undue emphasis to the conspicuous, newsworthy TNCs. Their role is more extensive and diverse than this.

A number of attributes are critical. The key characteristic of a TNC is that an enterprise controls assets, such as factories, mines, and sales offices, in two or more countries. Somewhat more narrowly, TNCs are usually defined as firms that control production in more than one country. A couple of ambiguous points should be noted here:

- The emphasis is on control, not just ownership of assets. Control means the exercise of management, and sometimes who is actively managing is difficult to know.[1]
- a firm does not have to be big to meet this criterion. Particularly in a world which now has hardly any national restrictions on the international movement of assets (more on that soon), owning assets in more than one country is not logistically difficult. For those of us in the island country of Australia, other nations are some distance away; producing in more than one country is logistically difficult. But this is much less true of countries which share a border, with populations mingling internationally.[2] But the term TNC has by convention been applied only to big companies, for which production in more than one country is a deliberate strategy rather than an accident of location.

On this basis, the United Nations estimates that in 1990 there were 35 000 TNCs with 150 000 foreign affiliates, and foreign-held assets of $US1.7 trillion (United Nations 1992).

TNCs and communications technology

Communications technology relates to our understanding of TNCs in two ways: communications technology is central to the internal organisation of TNCs; and the companies which produce communications technology are themselves often TNCs.

Companies producing in multiple countries and selling in many locations require mechanisms of internal communication, for accounting purposes and for managerial decision-making. One of the decisions a TNC must make is whether to run the corporation in a centralised way, with production decisions made from head office, or whether international plants (subsidiaries) should be administered autonomously. Another decision is whether the output produced by the company should be generalised (producing the same item in a number of countries, such as hamburgers) or specialised (different production plants produce different items—either components for a globally-produced item,[3] or specialist lines[4]). These decisions will be made differently in different companies and in different industries, but if there has been a general tendency in the last decade, it has been towards centralised control by head-office (at least regional centres) and globally integrated production.[5] This trend depends directly on complex intra-corporate coordination. This would not be possible in the absence of advances in communications technology, where efficient couriers, telephone, fax and computer links now make detailed international coordination feasible, and competitively necessary.

There are clear reasons why companies operating in communications technology are conspicuous among TNCs. While the invention of new technology is often undertaken by smaller, specialist firms, the commercial development of such technology requires that it be applied on as large a scale as possible. Over the past 100 years, and particularly over the past 20 years, the nature of technological development within the area of communications has made physically possible an increasing variety of forms of cross-national communication. For the companies which own the technology, it is often rational to become transnational.

The economic role of TNCs

Economic activity (trade, money lending and production) has been international since before there were nations (if that's possible!). In particular, trade has been attributed a central role in providing the wealth on which European capitalism developed. Critical here were companies like the giant English East India and Dutch East India Companies which amassed fortunes trading in exotic products such as tea and spices,[6] in the period prior to the rise of industrial capitalism in Europe. But these were not, at least initially, TNCs as defined above. These companies did not enter the new markets as bearers of capitalist industrial processes, but as traders—buying cheap and selling dear, and using their size and might to acquire local trading revenues, subordinating local political structures in the process.

Control over foreign production first emerged in tandem with colonial rule, with companies of imperial countries investing in assets in the colonies. But even this is not the activity of what we now understand as TNCs, for colonial investment remained under the aegis of the 'home' imperial state and thus, in a political sense even if not in a geographical sense, domestic.

TNCs as we know them are a late nineteenth and particularly twentieth-century phenomenon, where investment is not restricted to within the colony, but occurs between sovereign states. Their development is to be understood as a direct reflection of the development of capitalism from the late nineteenth century. A few factors stand out:

- The development of production technology, particularly mass production, meant that firms which possessed a technological advantage (either a superior technique of production or superior quality product) could sell their products profitably throughout the world. If transport costs are high, or prospective foreign markets are protected by import restrictions, it would be cheaper to produce in other countries than to export to those countries.

- Successful firms with access to finance could set up production or purchase productive assets in other countries as a form of corporate expansion. The development of public companies (companies which raise funds by selling off ownership through the stock market) in the late nineteenth century, and the concurrent growth of commercial banking, both provided companies with access to assets to fund growth.
- Two questions immediately follow: first, why would companies want to expand internationally rather than diversify within their country of origin; and second, what does this say about the broader nature of capitalist development. In the answer to the first question lies the terrain of the critique of TNCs; in the second, their justification.

For the first question, monopoly power, or, more generally, control over the market, is central.

Monopoly power of TNCs

Companies with sole ownership of some technology want to exploit the benefit that accrues from it as far and wide as possible. International production gives them a bigger market to dominate.

There are a couple of aspects here. First is the fact that any company with more advanced technology than its competitors is likely to be more profitable. It may be because its technology permits it to produce more cheaply; or, in the particular case of communications technology, because access to faster or better flows of information confers a competitive advantage. For example, the early use of mobile telephones at racetracks permitted gamblers to profit from the difference between on-course and off-course odds,[7] and the development of satellite geological surveys has given the larger mining companies which can afford them enormous advantage over companies relying on on-the-ground prospectors.

Another aspect is associated with the benefits of being big. Communication within a technological system, be it a telephone system or a computer system, invariably gravitates towards an international scale for two reasons. First, the technology is often very costly to set up. The development of computer systems or the launching of satellites is too expensive if it is for communication between a handful of agencies. The high start-up costs must be spread over as many users as possible. Companies in these industries spread internationally because it is cost-effective. Second, the more agents a communications system can communicate with, the more agents will want to join it. For example, a telephone system or a computer system which only permits domestic communication will have fewer subscribers than one which permits international communication. Thus there are clear technological reasons for some parts of the communications industry to operate on an international scale.

Companies with access to finance, irrespective of technological advantage, may be able to buy out rivals in other countries, and establish a monopoly position in their markets.

Within communications industries this most applies to the media, and most conspicuously, to newspapers. Not only do companies want to own newspapers in multiple countries because they 'know the business', but communications technology itself enables companies to 'network' copy, reducing their costs significantly compared with independent newspapers.

Companies using imported inputs may wish to control their supply of those inputs. For example a steel works may wish to own foreign iron ore mines, or a food processing company foreign farms. Within communications, the same tendency arises, although to a lesser extent than is found in relations between primary and secondary industry.[8]

Each of these rationales for internationalisation could be expected to be more profitable than to diversify into other industries in the home economy. What makes this profitability likely is that the rationales are associated with some form of market domination. This is compatible with the image of TNCs as power brokers and market manipulators. Whether that is a fair and universal depiction we will see shortly.

International expansion by TNCs

For the second question, the inherent tendency of capitalism to expand is central. This tendency rests on the proposition that unlike all previous epochs of history, capitalism is based on the competitive pursuit of profit. Competition creates the inducement for technological change and the growing size of firms. Technologically backward firms go out of business, and smaller firms are generally either less profitable or restricted to certain industries. This makes capitalism a growth-oriented system. That doesn't mean that it is in fact always growing, but periods without growth are also periods of economic crisis.

One dimension of the competitive, growth-oriented nature of capitalism is the spatial (international) extension of markets (trade) and finance (credit and investment) and, associated with them both, the international relocation of investment. TNCs participate in this process, and by controlling production in multiple countries, they are important in determining global patterns of investment and trade. But there is more to the process of the international extension of economic activity than TNCs.

The last twenty years have seen significant developments in this internationalisation, including in TNCs. Since the 1970s, the movement of money and commodities internationally has accelerated at an historically unprecedented rate. On a global scale, trade grew rapidly in the 1970s, at twice the rate of the 1950s and 1960s. But the international movement of money grew even more rapidly, for while trade increases levelled off in the 1980s, international money movement kept growing.[9] This was associated with the rise of debt-financing of investment, and the growth of international banking. It was facilitated by many countries lifting their restrictions on international money movement and floating their exchange rates. Economists now talk of a single, global money market which knows few or no national barriers.

This has been an environment in which TNCs expanded as the bearers of this internationalisation. Between 1985 and 1990 on a global scale, foreign investment by TNCs grew at an annual rate of 34 per cent, compared with international trade at 13 per cent and global production at 12 per cent (United Nations 1992). Large, and even middle-sized, companies were advanced by the international banks' long lines of credit to invest where they thought most profitable—in whichever industry and whichever country.

Australia is a clear and dramatic illustration of this trend. In the 1960s, there were but a handful of TNCs of Australian origin. But by the mid-1980s, virtually all large (even by Australian standards) Australian companies held international assets and engaged in production outside Australia.[10]

So TNCs are no longer the exception as a form of organisation of large corporations, but nor are they path-breakers in creating an internationally integrated economy. The critical change is in the underlying ordering of the international economy. The integration of global capital markets means that when we go to our local bank for a loan the bank sources that loan from an international financial system, whether it is itself a large international bank or not. The growth of trade means that we have greater access to the products of other countries—we both import and export more. This in turn means that it cannot be assumed that local markets will be supplied by local producers: they will be supplied from the cheapest international source.

This is where TNCs come in. They use their access to international credit to undertake production in the place (nation) where it is most profitable, and then export to markets where it is less profitable to produce. To a great extent, this adjustment would occur in the absence of TNCs: if it is profitable, someone will probably do it. But TNCs specialise in knowing where to do it, how to finance it and where to sell it. They are, in this way, just playing out the pursuit of profits, which defines capitalism.

Critiques of TNCs

As just posed, the role of TNCs sounds quite innocent. However, a number of criticisms are levelled at the activity of TNCs, which would suggest that there is more to their role than that.

TNCs exploit cheap labour in poor countries

As part of the international profit agenda of TNCs it is argued that, in the last twenty years, they have been taking industries from high wage countries to low wage countries. They have thereby 'deindustrialised' the richer countries (they have taken industry off-shore), and exploited cheap, non-unionised workers in bad working conditions in the poorer countries.

Poor countries, and the cheap labour they provide, are not high on the agenda of TNC investment strategies. Over 80 per cent of investment by TNCs is in 'rich' countries, and 70 per cent is investment between just three 'countries'—the United States, the European Community and Japan. Of the less than 20 per cent going to poorer countries, two-thirds goes to just ten countries (United Nations 1991).

Even so, there is TNC investment in poor countries, and there is some pattern of labour-intensive industrial production (for example, textiles production, assembly work) being relocated from advanced capitalist to poor countries, largely in response to the low level of wages in the poor countries.[11] This is indeed more profitable for TNCs in certain industries, but several points should be noted. First, for the workers so involved, the alternative to low-wage work for TNCs is invariably not high-wage work (the TNC would not have relocated if this were the case), but unemployment. Second, it is generally the case that TNCs pay higher wages than do the local employers. The real exploitation occurs in the smaller, 'backyard', locally owned firms. Third, it is argued that it is investment which has historically made the rich countries rich, and if poor countries had received more foreign investment, they would have become richer. If poor countries lack the local funds for investment, they must get them from outside and, despite the fact that profits accrue to foreign owners, they can expand their economies. Indeed, this is precisely the story of Australia: it became one of the wealthiest countries in the world on the basis of foreign investment.[12]

In general, the problem of the poor countries is not that they are exploited by TNCs, but that they lack development; or as it has been put, 'the problem of the poor countries is not that they have been exploited, but they have not been exploited enough'. The problem of poor workers in those countries is not that they are exploited by transnational capitalists, but that they are exploited by all capitalists.

TNCs dominate markets, and exercise monopoly power

The critique here is that the size of TNCs means that they can drive competitors out of national markets and establish monopolies, and so increase their profit at the expense of consumers and smaller producers.

Here, the critique is not of the trans-nationality of capital, but its bigness *per se*. Companies do not have to be transnational to be big. Nonetheless, to the extent that TNCs do exercise such power, it is the responsibility of nation-states to institute anti-monopoly controls; not of TNCs to 'self-regulate'. Perhaps, however, it is argued that TNCs are able to exercise power over national governments, and somehow prevent such anti-monopoly controls. This is a distinct critique, to which we now turn.

TNCs can dominate national governments

It is argued that TNCs are often so large, and have sufficient international influence, to be more powerful than the state institutions of the countries in which they invest. They can thereby extract favours and, at the limit, secure the overthrow of governments.

While such cases are by their nature difficult to prove, it is almost certain that TNCs have at times exerted decisive political impacts. There are well-documented cases of the role of TNCs being active in the overthrow of 'hostile' (usually socialist) governments, especially in Central and South America, and particularly in the 1950s. In part, this is an issue of TNCs as bearers of 'foreign' influence, an aspect looked at shortly.

In part also this critique is about TNCs as representative of capitalist class interests; interests they share with local capitalists, too. Much of the political power they exercise is therefore not just because they are transnational, but because all large property owners exercise significant sway over the state.

Their one distinctive source of power is due to the lack of an effective international system of state regulation. In its absence, TNCs can use the unregulated international environment to their advantage. Most conspicuous here is the capacity to minimise taxation payments by declaring their profits in countries with low tax rates and making no profits in high tax countries. This can be done by an accounting measure called 'transfer pricing'.[13]

The less dramatic, and no doubt more pervasive, issue here is the capacity of TNCs to extract favours such as taxation concessions, non-unionised labour, favoured access to resources etc. Yet here, as above, it is important not to exaggerate the favours acquired by foreign capitalists over the favours acquired by local capitalists. Where favours are acquired by bribery, for example, there is no reason to isolate the activities of TNCs.

Yet some favours, such as tax concessions are exclusively offered to TNCs. Why? Quite simply, they are offered by states desperate for investment, and these states must 'compete' internationally to attract mobile investment. More conspiratorially, it is argued that TNCs can, in concert, threaten to withdraw their investments if local government policies do not meet their interests. This is a form of power, but it does not constitute power over the state. Indeed, the very notion that TNCs can exert power over the state is based on a simplistic conception of power.

TNCs (all companies) have power, and states have power, but it is not the same power; they are not mutually exclusive. State power is required to secure social stability and sustain economic activity. TNCs (and all capitalists) need this power to be exercised. They may, and almost certainly do, seek to influence the way it is exercised, but this is a different matter from usurping state power. It is about recognising the capacity of the capitalist class (TNC and local) to see that the state sustains their interests. This, then, is a question of the role of the state in sustaining the dominance of the capitalist class; it is not one of TNCs over-riding the power of the nation-state. To focus on TNCs implies the false suggestion that, in the absence of TNCs, class divisions in society would not be a matter of concern.

TNCs undermine national sovereignty

The response above is, however, open to a further critique, even if it is recognised that TNCs cannot usurp state power. This is the proposition that TNCs, unlike 'local' corporations, serve foreign interests, and important decisions about the national economy are made in boardrooms in New York and Tokyo, not within the nation concerned. This has a few dimensions: TNCs take profits out of the country; the influence they seek to have in the formation of state policy is to serve their global interest, not the 'national interest'; and their impact breaks down national distinctiveness, economic, social and cultural. These latter two dimensions have been particularly raised in the context of communications technology.

There are many related issues here. For the sake of exposition economic and social aspects will be treated separately. Clearly, wherever there are TNCs, resources flow internationally. TNCs do take profit out of countries, but only when the local operation is profitable. The only way to avoid the outflow is to prevent the investment in the first place. But this involves a loss of economic activity. It may be that the profit outflow is worth the cost.

Beneath this 'profits versus economic activity' dilemma is a broader issue of whether 'foreign' companies should be cast as hostile, and 'local' companies as patriotic—whether decisions made in boardrooms in New York or Tokyo are really any different from decisions made in boardrooms in Sydney or Perth. If there ever was a difference, the issue has changed in the past decade, in two important ways.

First, with the growing number of TNCs, profits flow in many more directions. For example, while foreign TNCs take profits out of Australia,[14] Australian TNCs bring profits into Australia, though invariably to move them out again as new international investment. In this international era, there is no clear pattern of resources being drained from one country to another. It now cannot be easily argued, for example, that Australia is a client of the United States economy when, in 1988, Australian investment in the United States was greater than United States investment in Australia!

Second, with the growth of international money markets and the use of debt-financing by industry, most cross-national income flows are interest payments, not profits. This is significant because the loans which give rise to the interest payments are taken out voluntarily by local firms, predominantly for local investment, though also to fund international investment. So while income leaves the country, it can hardly be argued that it is associated with a loss of national sovereignty.

If this starts to sound confusing, it is because the relationship between nations and companies is becoming more complex: there is no clear 'us' and 'them' delineation of companies, and the wealth accumulated by companies does not directly flow to their nation of origin.

This points to the general issue of the notion of sovereignty. In an economic dimension, if investment, credit and commodities move freely in and out of countries, in what sense can we still talk of a discrete 'national economy'? The nation is no longer a clearly delineated economic unit. Ownership structures of industry are now too complex to have a clear differentiation of foreign and local ownership, and even when ownership is local, it could be by a local TNC, or be funded by foreign credit. Moreover, in a globalised economy, nation-states policies (despite what political leaders would like us to believe) are increasingly less able to affect national differences in economic activity. Sovereignty, at least in an economic sense, is (at best) an anachronism.

It follows that we must also think critically about the concept of a 'national interest'. If we cannot clearly distinguish a national economy, and the companies which run industries are increasingly international (not just 'foreign'), then whose interest is the national interest? It is argued, for example, that it is in the 'national interest' that industry should be locally owned. But when the share equity which denotes ownership is funded by international credit, or where 'locally owned' companies are doing a substantial portion of their investing internationally, there is not a clear difference between local and foreign ownership. Local ownership is not obviously intrinsically preferred. Certainly, there can be presumed no national 'loyalty' by locally owned companies: they invest where it is profitable. Of course, politicians and governments would have us believe that collective effort yields collective reward, but it is far from clear that there is a systematic process which ensures that the rewards are nationally distributed.

Perhaps, however, the issues of sovereignty and national interest become clearer outside the directly economic sphere. Here, the argument is that the loss of national sovereignty is more subtle and pervasive than is captured by economic analysis. This will be looked at specifically in relation to the impact of communications technology and TNCs.

The critical issue is that control over communications technology means control over information, and control of information is a precondition of political power. This has two dimensions: control over popular consumption and culture and control of national centres of security and intelligence.

The first relates to the capacity of TNCs to create a social environment which is receptive to the products of TNCs. This means, in the first instance, creating a demand for the mass consumer goods produced by the TNCs. Associated with this is the construction of a social environment in which such demand appears 'natural'. It is argued, for example, that foreign controlled media—from current affairs to sport to soap-operas—give societies, particularly outside the United States, a foreign-constructed image of politics and social relations. This is generally associated with broader explanations of United States dominance of the world economy. Local culture and values become subordinated to foreign culture and values. The world takes on American values of consumption and individualism. For example:

> The corporate takeover of (popular) culture for marketing and ideological control is not a patented American practice . . . [but] . . . is carried to its fullest development in the United States. Cultural–recreational activity is now the very active site for spreading the transnational message, especially in professional sports . . .
>
> Major sports are now transmitted by satellite to global audiences. The commercial messages accompanying the broadcast . . . and often worn on the uniforms of the athletes constitute a concerted assault of corporate marketing values on global consciousness (Schiller 1991, pp. 23–4).

The issue of cultural imperialism is discussed by Lelia Green in chapter 12. Let us here concentrate on its relationship with TNCs and sovereignty. It is clear that the trend towards global information systems and global media is a reality. But the unique role of TNCs in this process should not be exaggerated. It is limited to a couple of factors. One is that the involvement of TNCs often means that foreign sources control the flow of information. This issue cuts two ways. The negative side is the possibility of news being converted into international propaganda, as found in the United States media's coverage of the invasion of Iraq in 1991. The positive side is found in overcoming domestic propaganda and censorship of information, as seen dramatically in the capacity of satellite-linked media to transmit to the world pictures of the massacre in China's Tiananmen Square in 1989.

The other factor is that the creation of globalised communications occurs under the control of a relatively small number of large companies, often because the cost structure of the industries concerned means that only large producers can operate profitably. That is far more restricted a criticism than is commonly found, as in Schiller above, where TNCs are attributed the role of creating cultural hegemony, 'Americanising' the world.

Let us leave to one side the moral critique that American values are intrinsically bad for non-Americans. (It is a judgment that should not be presumed.) TNCs themselves are not to be held responsible for the fact that the world's television and cinema screens are dominated by American content. Indeed, this phenomenon has little to do with undertaking production in multiple countries (which, it will be recalled, is a characteristic of TNCs). These programs are exported as commodities, just like wheat or shoes, and they are imported into other countries because their buyers believe it will be profitable to transmit them locally.

The concern which follows from the moral critique, therefore, is not just that the supply of foreign goods and information is a threat to national political and cultural sovereignty, but that there exists a real local demand for these items; a demand which cannot be dismissed by the somewhat patronising proposition that TNCs can make people want things they don't really want.

The second sovereignty issue relates to the way in which centralised, internationalised communications technology can be used for foreign control over national security. Foreign TNC control of satellite and other telecommunications technology provides the means for elaborate external surveillance, which has profound political and military implications. There is ample scope for this, both in the illicit form of spying by means of extra-territorial surveillance, and by the control over the information systems on which national security is operated. Indeed, it is argued that much satellite technology does not just have military application, it was developed for military purposes (see, for example, Matellart 1979). The use of satellites for weather forecasting and geological surveying was first developed by military-funded research. Much of the recent development of satellite communications technology is attributable to the United States' 'Star Wars' military program.

This issue of the link between TNCs and military applications is perhaps the most disturbing of all. It has been a growing concern since President Eisenhower warned of the emerging 'military industrial complex' in the 1950s. But the concern must be clearly targeted. The production of defence technology is only profitable because governments heavily subsidise it. The fact that the companies which produce the technology are invariably TNCs does not mean that loss of national sovereignty by the application of such technology is the responsibility of the TNCs, though they may well find the whole exercise highly profitable. The problem comes back not to the subversion of nation-states by TNCs, but the primacy given by all nation-states to the needs of all capital.

Conclusion

TNCs have played a significant role in the recent expansion of the communications industry internationally. They have been central in developing communications technology, and, by use of the technology itself, have spread rapidly the application of technological innovation. The effect has been to break down national barriers, both economically and culturally, though not, it must be emphasised, so as to eradicate national differences.

A consideration of whether domestic commodities, companies and culture are inherently preferred to foreign ones is beyond the scope of the current analysis. But a number of issues have been discussed above which reflect directly on that broader question. There is a widespread tendency to focus much blame for negative experiences of the international economy on TNCs. They are often seen as manipulating economic and political processes to their own advantage, and constructing the world in their own image. They are in some sense immoral.

But perhaps they are just good capitalists. With the use of communications technology applied globally, they know where and how to raise funds; they know where it is best to produce; they know how to produce profitably; and they know where and how to market their output. Most of all, they know that to be big is to survive.

The evaluations are not incompatible, but they lead to quite different consequences. The moral critique suggests that the world will be improved by the restriction or even eradication of TNCs. Should this happen, it would be argued, nations will be sovereign, and the will of the people will be heard. The 'good capitalist' critique says not. TNCs are a reflection of globalised accumulation, and in the absence of TNCs something approximating the current system would be created by means of trade and national replication of foreign technology and foreign products. Production under the control of local capitalists is no more the 'will of the people' or an expression of 'the national interest' than is production by TNCs. It is done with the same objective, though probably less skill, and with similar outcomes.

If there is a concern with the current world impact of communications technology, and the exclusivity of its control, the underlying issue is not so much that it is controlled by TNCs, but that it is controlled internationally and within nations by a minority, property-owning class. While there may be conflicts within that class between TNCs and local capitalists, these are trifling compared with the combined power of this class to determine the lives of ordinary people throughout the world.

Endnotes

1   In the 1991 takeover battle for Fairfax newspapers, the question of who would be managing depended upon interpretation of just how much influence Kerry Packer would exercise. But influence is hard to measure.

2   In Europe, for example, someone could run two cake shops in adjoining towns, but one could be in Germany, the other in Austria; hence a 'transnational corporation'.

3   For example, IBM undertakes its research and development in multiple countries, relying on intimate coordination between its world-wide laboratories (McCann 1986).

4   An example here is a food company producing its global supply of canned tomatoes in Italy and its global supply of frozen peas in Australia.

5   The decision of how to organise a TNC on an international scale is an important regional question. In particular, countries and cities vie to become regional centres for transnational industries—for example, Sydney's attempt to be recognised as financial centre of the South-East Asian region. On the significance of these issues, see, for example, Langdale 1989.

6   Notice that for all their wealth and power, these trading companies do not automatically fit the criteria of TNCs, for trading *per se* does not require control of assets in other countries.

7   It will be recalled that in Frank Hardy's *Power without glory*, a fictionalised biography of Melbourne businessman John Wren, the leading character made his start in business as an SP bookie, using homing pigeons to secure race results before they arrived at radio stations by telegraph. He was thereby able to adjust his book before the results became 'public'. There is a long history of those having superior communications technology being able to exercise it for market advantage (see Carey 1986).

8   Nonetheless, there is an observable trend for producers to want to control distributors, and for distributors to want to control retailers. For example, the conspicuous link in ownership between film distributors and cinemas means that the distributors determine the films shown to the public.

9   Between 1972 and 1985 funds raised in international money markets increased at 23 per cent per annum, while trade grew at 13 per cent

per annum, itself a significant increase over the 1950s and 1960s (United Nations 1989, p. 64).

10 By 1989 fourteen of Australia's fifteen largest companies held more than 25 per cent of their assets outside Australia; four of them held more than 50 per cent. Of Australia's largest 100 companies, only fifteen have fewer than ten per cent of their assets held overseas (Thomas 1989).

11 It is important not to assume that wages are the sole nor always the primary criterion of where to produce. If that were the case, all production would occur in the Horn of Africa. Other factors are the productivity of labour, the infrastructure provided in a country, proximity to markets, as well as the 'deals' which can be struck with individual national governments.

12 Clearly there is some debate as to whether that was a desirable path— indigenous people make a case that it was not. The point here is not to insist that capitalist development is good, but that poor countries are judged poor by the standards of developed capitalist countries.

13 Transfer pricing arises because TNCs continually shift funds and assets internationally within the company. The prices they put on these funds and assets are only for internal book-keeping purposes; they do not necessarily reflect market prices. Companies can construct these (internal) prices to shift income and so make branches of their operations look more or less profitable than they 'really' are, and thereby minimise global taxation payment.

14 It should be noted that not all profits of foreign companies are repatriated. Some portion stays to fund new investment. It is important not to have the image that TNCs invest and then withdraw. They tend to keep on investing for the same reason that they started investing— because it is profitable.

References

Carey, J. 1989, *Communication as culture: essays on media and society*, Unwin Hyman, Boston

Langdale, John 1989, 'The geography of international business telecommunications', *Annals of the association of American geographers*, vol. 79, no. 4

Mattelart, A. 1979, *Multinational corporations and the control of culture*, The Harvester Press, Brighton

McCann, J. 1986, 'Global management systems for multinational corporations', *Towards a law of global communications networks*, ed. A. Branscombe, Longman, New York

Schiller, Herbert 1991, 'Not yet the post-industrial era', *Critical studies in mass communication*, vol. 8 no. 1, March, pp. 13–28

Thomas, Tony 1989, 'Taking on the world', *Business review weekly*, 24 November

United Nations 1989, *Foreign direct investment and transnational corporations in services*, United Nations, New York

——1991, *World development report 1991*, United Nations & Oxford University Press, New York and Oxford

——1992, *World investment report 1992*, United Nations, New York

United Nations Centre on Transnational Corporations 1978, *Transnational corporations in world development: a re-examination*, United Nations, New York

Annotated bibliography

Crough, Greg and Wheelwright, Ted 1982, *Australia: a client state*, Penguin, Ringwood

An Australian application of critical analysis, though not specific to communications technology.

Radice, Hugo 1975, *International firms and modern imperialism*, Penguin, Harmondsworth

The interchange between Bill Warren and Robin Murray, reproduced in this collection, is a useful introduction to the debate about the relation between transnational corporations and nation-states.

Radice, Hugo 1984, 'The National Economy: A Keynesian Myth?', *Capital and class*, vol. 24

Jenkins, Rhys 1987, *Transnational corporations and uneven development*, Methuen, London

Radice and Jenkins provide excellent responses to the radical critique of TNCs, focusing instead on the internationalisation of class relations.

Schiller, Herbert 1981, *Who knows? Information in the age of the Fortune 500*, Ablex, Norwood New Jersey

Schiller, Herbert 1991, 'Not yet the post-industrial era', *Critical studies in mass communication*, vol. 8, no. 1, March, pp. 13–28

Herbert Schiller has written extensively, offering radical critiques of the role of TNCs in the information and cultural industries.

12

MISSING THE POST(MODERN): CORES, PERIPHERIES AND GLOBALISATION
Lelia Green

Is it possible to identify the town square at the heart of the global village? In a networked world is there still a core and the periphery—or are all places equally enmeshed and interdependent?

Core/periphery theory suggests a usefulness, within each particular sphere, for distinctions of have and have-not. It addresses the issue that nations, communities, individuals are linked by relationships of power and dependency which vary according to the specific circumstances of the situation considered. I will be arguing that core/periphery theory is best related to aspects of our life which might be called 'modern' (with a specific usage of that word) and industrial.

Two of the key theorists to work in this field are Immanuel Wallerstein with his global systems theory, and Harold Innis. Innis' book *Empire and communications* (1950) analyses the role of communication in defining the centre and the margin, and the relative dependencies which result. For nearly 50 years core and periphery concepts have served technological theory, and communications theory, well.

This chapter examines ways of seeing the centre and the margin, especially as they concern the third world and cultural imperialism. Offering alternatives to simple oppositions it goes on to look at issues of gender—female/male distinctions remain a cross-cultural core/periphery demarcation.

Australians and New Zealanders live in societies which have been characterised as post-colonial and post-industrial. Post-coloniality identifies a past (and present?) peripheralisation. Post-industrial refers to the status of information societies (Bell 1973). These societies also include many aspects referred to as postmodern. This chapter asks—what is in the post? Are these new organising principles required to examine social relationships and power between different groups of people?

Technology and society in the global village

There is no doubting the relevance of certain places to people's lives. At the neighbourhood level, the community focuses its common life upon certain buildings and services. These meeting places include shopping malls, schools, hospitals, cinemas, libraries and shire halls. At a state level, policy formation and decision-making centre on the capital city—other areas of the state become peripheral, linked to the core via communication technologies. Nationally, Australia has a number of cores: Canberra is the political capital, Sydney the business capital, and Melbourne the centre of high culture and the arts—including gourmet food and Victorian architecture.

According to some accounts, core/periphery distinctions resemble a set of Russian dolls in which each figure contains a smaller figure at its core and an element of peripheral space. Yet with Australia the area of life considered—the context (economic, political, cultural)—decides the core. Areas do not neatly overlap, each enclosed within another. Further, there is an ambiguity as to the direction of the power relationship. As Angus and Shoesmith (1993) point out, 'Innis has provided us with . . . an understanding that centres are as much dependent upon their margins as the margins are on the centre'. If the core is dependent upon the periphery, can the core be the determining factor? Does the periphery determine the core?

Deconstructive philosophies and discourse analysis—postmodern tools—address these concerns. Ann Game, in her book, *Undoing the social: towards a deconstructive sociology*, starts with 'the basic semiotic assumption that culture or the social is written, that there is no extra-discursive real outside cultural systems' (Game 1991, p. 4). In other words, the way people look at core/periphery issues creates the text which is analysed. There is no 'real' core and periphery apart from what is seen that way.

Questions of 'reality' intensify with internationalisation—a process by which nation-states become increasingly interdependent. Yet cores and peripheries continue to be determined by context. In currency, for example, Australia accepts the United States as its core reference point—the Australian dollar is most likely to be quoted in terms of the US dollar. Australia's democratic and administrative infrastructure, however, remains closely modelled on the Westminster system (Britain's most abiding export of the eighteenth and nineteenth centuries). Discussions of trade and exports, foreign policy and free-market trading partners, increasingly concentrate upon the nations of South-East Asia: our nearest neighbours, and the fastest growing region of the world economically.

In globalised systems, even the *desire* to construct core/periphery distinctions can be thwarted; for example, the contest between the Yen, the US dollar and the Deutsche Mark to be the core world currency. Can any country control the value of its currency on the foreign exchange markets? National financiers may try—by changing interest base rates, or by buying or selling their currency—to affect the price, but the money market is too big to be ruled by such measures. Many authorities believe that power is now vested in the market itself, beyond the control of any particular nation.

No longer distinguishable, cores and peripheries link together forming global networks.

As early as 1968, Marshall McLuhan and Quentin Fiore commented that 'Today, electronics and automation make mandatory that everybody adjust to the vast global environment as if it were his little home town' (1968, p. 11). Communications media return to us images of village-like encounters, but on a global scale; with Bill Clinton more easily recognisable than a neighbour, and Bart Simpson more endearing than any local child.

In the developed world new communications media create an environment founded upon electronics. People are no longer members of social networks according to proximity alone. The 'psychological neighbourhood' permitted by technologies such as the telephone and E-mail allow individuals to keep in touch with those who might be hundreds or thousands of kilometres away (at a price). Microelectronic pervasiveness is accelerated by digitisation (which allows information to be standardised for multiple applications and transmissions) and by miniaturisation (which allows the incorporation of electronic components into more and more areas of life).

Whatever is considered: international phone calls, transnational corporations, the communications systems of the Gulf War, the power of information networks is to relocate human experience in a global arena. At one time news of a murder, like two-year-old James Bulger's, could terrorise a community—nowadays such stories terrorise a world. James' abduction from a British shopping centre by two 10-year-old boys was recorded on video by security cameras, and broadcast as international news. In communities linked by the image of the child led to his murder, anxiety and tension rose. It was as if he were led away time and time again, in thousands of towns and cities; not once, in Bootle. The experience of loss, sorrow and horror was relocated globally. Yet the audience could do nothing practical or personal—they could not cook meals for the family or leave flowers on the railway bank where James' body was found. They experienced the emotional power without the release of a response (Steenbergen 1993).

The technology captured, distilled and reproduced the moment of the abduction, concentrating the visual impact by simultaneously voicing the child's eventual death. The emotional essence of the event was stripped from any context other than that which heightened the effect upon the audience, and exported to the world. The commentary offered was one congruent with postmodernity: random, senseless, arbitrary, fragmented. Very little coverage added that Britain has 'one of the lowest murder rates in the world' (Kettle & Phillips 1993, p. 2), or that the probability of an under-five in that country dying at the hands of a stranger is only slightly higher than the risk of spontaneous human combustion. In a country of 55 million people an average of one child a year is killed in this way.

While packaging and commodifying the moment when James was led away, the technology lacked the resolution to allow easy identification of the abductors:

> The fact that it is all recorded on a security camera makes it even worse. Powerlessly we see the tragedy unfold through a medium installed ostensibly to protect the innocent from criminal wrongdoing. We are therefore doubly affronted, both by being made complicit in this terrible tragedy and by the demonstrable fact that such 'security' devices are clearly anything but (Kettle & Phillips 1993, p. 2).

> We are all totally frustrated . . . Why are those video pictures so crap? Why didn't they go into Dixons [a British chain of electronics shops] and buy one of those home video machines? (Mrs Connolly, cited in O'Kane 1993, p. 21)

Neighbourless electronic neighbourhoods, the trading of vital information for capital gain, the packaging and sale of images—all result from the commodification of individual lives. This disembodiment of the communication from the communicant, of information from the informant, is sometimes seen as the prime constituent of the postmodern, and of post-industrial society.

Postmodernism, peasants and the proletariat

For much of human history (as James Carey 1989, pp. 203–4 reminds us), communication travelled at the same pace as physical transport. A letter, or an order, sped as fast as the human carrier could; by horse for example, or by boat. There were exceptions. The network of bonfires throughout England at the time of the Armada were intended to signal invasion by the Spanish. Similarly, Native Americans used line-of-sight smoke signals, while African drumming relied on aural range.

These communication patterns characterised agricultural societies where a majority of people worked on the land. In such communities most important communications took place between individuals and their immediate social network; there was little need to send messages outside. Only a small proportion of people were geographically mobile, and travel for any distance involved danger and difficulty. Although messages might enter the community, their relevance was determined by the ability of the communicator to enforce the communication. A call to arms was always more serious when it was accompanied by a press-gang and/or a hanging judge to deal with 'deserters'.

Industrial cities brought a completely new challenge in terms of organisation and commmunication. Subsistence-farming communities, whose notional surplus had been used for the enrichment of the church or the gentry, were caught up in changes associated with the creation of a new urban market. The accompanying agricultural revolution allowed the hundreds of thousands in the cities to be fed by a much reduced population working the land. The flow of primary products from the periphery to the core, and of manufactured goods from the core to the periphery, characterises industrial societies, and apparently supports core/periphery theory.

Since new technologies—rail services and the electric telegraph—allowed communication systems to be divorced from the limitations of living carriers, people no longer needed to transact business face to face (Carey 1989). Trading, and the taking of profit, became an impersonal activity. Much of core/periphery theory embodies a sense of manipulation of the periphery by the rapacious agencies of the core, for financial advantage. Consequently a critical (radical left) perspective characterises much of the analysis. The transition from agricultural production to a modern, industrial, society is seen in cataclysmic terms:

> All revolutions are epoch-making that act as levers for the capitalist class in the course of its formation; but this is true above all for those moments when great masses of men are suddenly and forcibly torn from their means of subsistence and hurled onto the labour market as free, unprotected and rightless proletarians. The expropriation of the agricultural producer, of the peasant, from the soil is the basis of the whole process (Marx 1976, pp. 875–6).

Compared with the romantic reinterpretation of peasant life as harmonious, communal and natural, modernist perspectives communicate negativity and loss. Identifying Edvard Munch's image *The Scream* as quintessentially modernist, Frederic Jameson comments that the modern involves the 'alienation of the subject' (1984, p. 63). Modernist texts include the early industrial city, mass production, mass broadcast communications and the theories of Marx and Freud: individual *angst*.

Postmodernism, on the other hand, displaces the alienation of the subject 'by the fragmentation of the subject' (Jameson 1984, p. 63). A filmic illustration might be a comparison of the modernist *1984* with the postmodern *Brazil*. Postmodernism fragments both the subject and the self. According to Jameson it means:

> the end for example of style, in the sense of the unique and the personal, the end of the distinctive individual brushstroke (as symbolised by the emergent primacy of mechanical reproduction). As for expression and feelings or emotions, the liberation, in contemporary society, from the older *anomie* of the centred subject may also mean, not merely a liberation from anxiety, but a liberation from every other kind of feeling as well, since there is no longer a self present to do the feeling (Jameson 1984, p. 64).

The postmodern promotes surface rather than modernist depth, commodification rather than exchange, and consumption rather than production. Put crudely it reduces human society to its icons and to its material possessions. Communities no longer need members, they need consumers. Jameson, comparing *The Scream* with Andy Warhol's (postmodern) images, comments that Warhol's work:

> turns centrally around commodification, and the great billboard images of the Coca-Cola bottle or the Campbell's Soup Can, which explicitly foreground the commodity fetishism of a transition to late capital, *ought* to be powerful and critical political statements (Jameson 1984, p. 60).

Jameson perceives the dissociation of form from content as a political development. In describing postmodernism he does not celebrate it. Surfacing is not necessarily *good*, it simply *is*. Postmodern analysis lacks the political up-frontness of tortured core/periphery commentaries, yet the politics behind postmodern theories remain oppositional to the interests of those elites which benefit financially from the promotion of consumption.

To concentrate on a political underpinning of postmodernism, however, is to frustrate the utility of a surface analysis. Fragmentation and commodification are appropriate conceptual frameworks for examining the technological realities of mobile phones, global networks, transnational capitalism and international markets. Fragments and surfaces in many respects characterise the technologies, and the lives, of people today. There is a positive aspect to fragments—it is easier to integrate fragments than wholes.

The polarisation which marks a modern society, with all its core/periphery overtones, is particularly obvious in the spheres of work and gender relations. The housewife—that modern invention of the late nineteenth and twentieth centuries—was socialised to espouse a house because the productive work of men (in the factories and offices) would otherwise be compromised by all the messy jobs of daily life (Oakley, 1974). Never before industrialisation had society as a whole been rich enough to consign young women exclusively to home duties. In pre-industrial societies child-rearing past toddlerhood was the traditional role of grandparents (mainly grandmothers, but with grandfathers duly teaching masculine behaviours). Young mothers (carrying their nursing babies) worked in cultivation and cottage industries, usually cooperatively, assisted by older children. Men spent their time hunting, gambling, building, trading and herding. Occupational divisions, although gendered, lacked the isolation, specialisation and rigidity which characterise work in the modern world.

The economic prosperity of industrialisation created a society in which a man was paid enough to support him and his family, and in which a woman resigned her job upon marriage. The 'career girl' could not also be a wife or mother. The distinctions would not tolerate ambiguity. The single mother gave her child up for adoption, the unhappy marriage rarely ended in divorce. Category boundaries were only negotiable by the immensely strong, or the immensely wealthy. (Usually people had to be both.) Industrial society is marked by the litter of lives of people who do not want—or cannot fit—such rigid compartments.

In Australia in the 1990s, a post-industrial society, the concept of work is one of the most abiding bastions of modernity. Why is it that faced with 10 per cent unemployment, and increasing mechanisation, politicians (mainly male) continue to talk about ways of creating 'new jobs'? Why is there so much talk about stress, and so little talk about integrating fragmented lives by encouraging teleworking or more *secure* part-time work? A postmodern approach might be to challenge the given of the 8-hour day and the 5-day week. If society guaranteed the rights of people negotiating their own working hours, or the number of weeks they work in a year, or the right to take unpaid leave, then greater flexibility in the workplace could help the 90 per cent of jobs stretch to the 100 per cent of people who want them. It might also allow for a more flexible, responsive, needs-centred society. Technology could help in this:

> The lack of boundaries both in hunting and gathering and in electronic societies leads to many striking parallels. Of all known societal types before our own, hunting and gathering societies have tended to be the most egalitarian in terms of the roles of males and females, children and adults, and leaders and followers . . . play and work often take place in the same sphere and involve similar activities . . . work and play have begun to merge in our electronic age. Both children and adults now spend many hours a week staring at video monitors (Meyrowitz 1985, pp. 315–6).

The thoroughly modern first, second and third world

The duality of core/periphery theory involves oppositional distinctions between places or cultures which may be seen as 'peripheral' and those at the core. Cores benefit at the expense of their peripheries. There is otherness, and self with the self conceived as positive to the negatives of otherness (whether or not the self is at the periphery).

Colin McArthur (1985) discusses the creation of the periphery (Scottish people—*homo celticus*) by the core (English/French society—*homo oeconomicus*) in terms of binary oppositions: 'Oppressed people the world over know this discourse to their bitter cost . . . What is important is that the Celt (or African, or Polynesian, etc.) is allocated his/her place, is constructed, in a discourse enunciated elsewhere' (1985, p. 65). McArthur's oppositions are followed and adapted here to compare conceptions of *homo informaticus* (first world) and *homo incommunicaticus* (third world).

Homo informaticus	*Homo incommunicaticus*
urban	rural
civilised	wild
rational	emotional
Christian	nonbeliever
astute	gullible
innovative	traditional
rich	poor
generous	needy

This duality is not offered as a 'real' explanation of differences between the first and third worlds, but as a demonstration of how a periphery can be constucted in opposition to the core. Such a construction demonstrates the ignorance caused by an economic bias—the third world receives scant attention from the first except as a potential market or as the recipient of aid. The information poor and the information rich rarely communicate, so ignorance is perpetuated.

People in the third world have less information, and less access to information technologies:

> The data listed below were gathered by UNESCO (1989); the base year for these figures is 1984 . . .

> • Of the world's 700 million telephones, 75% can be found in the 9 richest countries. The poor countries possess less than 10%, and in most rural areas there is less than one telephone for every 1,000 people . . .

> • In the United States a daily newspaper enjoys a circulation of about 268 copies per 1,000 people; in Japan the comparable figure is 562. The African average is 16.6 copies per 1,000.

> • Europe produces an average of 12,000 new book titles every year. African nations produce fewer than 350. Europe has an average of 1,400 public libraries per country where the public has free access to information. African countries have an average of 18 libraries per country (Hamelink 1990, p. 219).

Technology intensifies the disparity between information rich and information poor. The west lays fibre optic cables, introduces broadband ISDN, and develops international mobile phone links. When comparative disadvantage is considered in terms of telephones, newspapers and books, what hope is there for bridging the information gulf?

Wambi, in Brizzio (1990) identifies a second problem: 'Technology is like a genetic material. It is encoded with the characteristics of the society which developed it, and it tries to reproduce that society.' Information societies are post-industrial societies; their wealth relies not on industrial labour but in the processing, packaging and use of data. Information markets and related activities—such as teaching, law and the media—dominate post-industrial economies. Their technologies demonstrate the importance of communications. Is this social organisation a desirable model for the third world? Is it possible to create an information society which avoids the problems of the first world?

Even if the third world could equalise technological resources and technology use with the first, it would not necessarily wish to do so. Given that the Scots resent being constructed by the English then *homo incommunicaticus* would resist being modelled by, or in the negative image of, *homo informaticus*. Yet a result of the core/periphery mind-set is the dynamic of have/have not—it perpetuates generalisations and sidesteps the possibility of developing 'equal and individual' alternatives.

This either/or is a genuinely modern problem—if the information gap continues, the first world is (morally) untenable: if the gap is filled, the third world is (effectively) engulfed. There is a compartmentalised division between the core and the periphery. The boundaries are hard to blur, except at the price of difference itself.

Further, the core/periphery duality is inadequate for conceptualising power disparities, or interdependencies, in networked systems. An appearance of integration is underlined by the dislocation and fragmentation within and between societies which is the common experience of individuals living in them. Boundaries are illusory. There are elements of information poverty within all first world countries, and tranches of information wealth within the third world. There were capitalists in the second (Soviet Bloc) world and revolutionary socialists in the first. A theory appropriate to the analysis of post-industrial life—networked and excluded—would take account of this fragmentation. Postmodernism rules OK.

McKenzie Wark, analysing the end of the Berlin Wall, uses postmodern analysis to contrast the ('equally real') *territory*—or physical spaces of Berlin, and the *psychological map*. According to Wark, people use both physical and mental schema to locate themselves and form their sense of place:

> Where territory is filled up by people and their interactions, the map is composed of broadcast areas, satellite footprints, telephone networks, and the signs and images which accumulate through interactions in this abstract place-less space . . . In the territory people know where they are because they have roots there. On the map people know where they are by tuning into it: here we no longer have roots, we have aerials (Wark 1990, pp. 36–7).

Indeed the aerials of East Germany, receiving broadcasts which the authorities were powerless to block, were an important destabilising influence. Psychologically people had stormed the Berlin Wall even before the first unauthorised footstep into no-man's land.

Modernist, industrial verities of core/periphery theory—endorsing such geographical certainties as the Iron Curtain and the Free West—give way to the postmodern. A psychological mesh without a centre encircles a technological network without a core in a globe which has no periphery. The one truth offered globally is that there is no one truth, everything is negotiable and all perspectives are relative to the discourse through which they are constructed.

This is not to claim that core/periphery theories are entirely bankrupt: they have their uses. In an analysis constructed upon the dichotomy of have/have not—upon comparisons—modernism is an appropriate discourse. There is no doubting the accelerating impoverishment of the majority of the third world by the action of economic systems which arise in, and enrich, the first world. To talk about the first world, however, only in terms of core/periphery does not help create understanding of the new realities of networked culture—for those who are part of the net. Where people try to analyse the nature of the information, post-industrial, late capitalist society, postmodern analyses offer fruitful ways forward.

Speaking American and the globalisation of culture

The historical manifestation of empire was prochronistic (the opposite of anachronistic): it anticipated the structural divisions of modern, industrial societies. Empires depend upon soldiery, specialised administrators and rigid divisions of labour. By the time imperial theory and practice had become incorporated into the British Empire, the institution was undoubtedly modernist and indisputably organised according to a core/periphery model. Imperialism became the *bête noire* of critical theorists—and of millions of oppressed peoples.

Herbert Schiller, an American academic, argues that whereas people theoretically live in a post-imperialist era, the impetus of imperialism is served by the empires of vast transnational corporations. A colonising media communicates an 'all-encompassing cultural package':

> the English language itself, shopping in American-styled malls . . . the music of internationally publicised performers, following newsagency reports or watching the Cable News Network in scores of foreign locales, reading translations of commercial best-sellers, and eating in franchised fast food restaurants around the world . . . The domination that exists today, though still bearing a marked American imprint, is better understood as *transnational corporate cultural domination* (Schiller 1991, p. 15).

Cultural imperialism serves the economic and industrial ends of transnational corporate power by fostering an environment which promotes consumerism. Media products, portrayed lifestyles and marketed goods are all consumed by the colonised culture. According to this core/periphery model, globalisation—which has been defined by Featherstone as 'the increase in the number of international agencies and institutions, the increasing global forms of communication . . . the development of standard notions of citizenship, rights and conception of humankind' (Featherstone 1990, p. 6)—offers nothing new, just more of the same. The haves are exporting to the have nots, at a profit. Global communications are dominated by American multinationals, valorising the American way of life. Standard notions of citizenship are essentially American in nature. To be American is to be at the centre of global culture: the modernist Empire is dead, but core/periphery distinctions strike back.

This analysis is stuck in a non-digital groove. Apart from bemoaning the situation there are few alternatives on offer from the core/periphery mind-set. The postmodern response goes beyond post-imperialism to examine the appropriation of cultural communication by the recipient, rather than the power disparity between producer and receiver. There is more to be discussed than relative disadvantage. The analysis of absence is likely to miss the presence of alternatives. Nobody doubts the pre-eminence of American transnationals in cultural production on a global scale: the question is whether this amounts to mass Americanisation.

Worldwide audiences are not mindlessly receptive of American cultural products. The image of a hypodermic injection of media messages into the psyche of passive consumers has been repeatedly discredited. Are core/periphery theorists guilty of culturalist assumptions—believing that other cultures will necessarily crumble beneath the American onslaught? The postmodern interpretation of media effects involves a call to ethnography— the study of 'the irreducible dynamic complexity of cultural practices and experiences' (Ang 1991a, p. 161).

Objective analysis of television viewers tends to concentrate, as audience measurement strategies do, upon gross data such as whether a television set is on, which channel it is tuned to, and the number of people watching. It is a boundary-confined analysis: on/off, core/periphery. At this level of analysis global culture is American because United States corporations produce and market the lion's share of global media products. The subjective level, however, interprets the on/off as having a more subtle meaning. The fragmentary nature of fine-grained ethnographic observation paradoxically allows a reconception of the whole. Technology is less an instrument of cultural domination and more a medium of expression. Ien Ang offers this example of Hermann Bausinger's (1984) ethnographic research:

'Early in the evening we watch very little TV. Only when my husband is in a real rage. He comes home, hardly says anything and switches on the TV.' Here, comments Bausinger, 'watching television' has a very particular meaning, profoundly immersed in 'the specific semantic of everyday': 'pushing the button doesn't signify "I would like to watch this", but rather, "I would like to hear and see nothing"' (Ang 1991a, p. 161).

These observations make sense from individual experience of family life. The fragmentary nature of postmodern analysis does not prevent an overview. Understanding of such closely observed data is achieved through a series of disconnected, jump-cut images. The oppositional nature of the audience, their ability to interpret in their own way the programming they view, runs counter to the soft-imperialist claims of the industrialised, core/periphery models. Further, the Australian and New Zealand experience is that cultures do survive (sometimes through resentment) the disproportionate influence of other powers—British or American. We may watch 'The Cosby Show', we do not live it.

Ang asserts that the globalisation of culture:

should not be conceived as a process of straightforward homogenisation, but rather as a checkered process of systematic integration in which local cultures lose their autonomous sovereignty and become thoroughly interdependent, relying for their active continuation precisely on the appropriation of global flows of mass-mediated forms and technologies (Ang 1991b, p. 5).

Ang suggests, for example, that the 'curry eastern' is an appropriation by Indian cinema of the conventions of the American spaghetti western, while the burgeoning popularity of martial arts movies for Cantonese audiences incorporates and transforms 'James Bond style film narratives by using fists and martial arts as weapons, as well as drawing on traditional Cantonese values' (Ang 1991b, p. 6).

At one level this involves the assertion that the more globalised the communication, the more specific is the local interpretation; the more homogenous the product, the more fragmented are the meanings taken from it. The dominant reading (the intended or obvious interpretation of the media text from the point of view of the producers) is resisted or subverted. Postmodernism invokes its own duality of oppositions to discuss these issues, addressing globalisation/localisation, and homogenisation/ fragmentation. A rich, deep usage of the postmodern perspective is available through concurrent consideration of homogeneity and fragmentation—the global in the local and the local in the global.

Postmodernism, therefore, offers a valuable perpective for understanding the complex nature of the information society. Many attempts to ground postmodern theory in the lives of individuals are frustrated, however, by the need to generalise, to express an overview. The postmodernist is uncomfortable with the notion of betraying the fragmentary by privileging the integrated. The postmodern theorist holds onto the (fragmentary) core at the expense of the (integrated) periphery—avoiding a return to the dichotomies of modernity. The possible way forward however, is grounded in feminist theory—a celebration of the principal of equal *and* individual.

Women are arguably the group in western culture most versed in the politics of resistance and renegotiation, and the women's movement has achieved some success in the fight against the peripheralisation of women's issues. What women want—in the researching of women and advertising, in the case of the example to be quoted—may illuminate an approach which would allow more research of the global within the context of an agenda set by the local. Not surprisingly, Chris Adams rejects the core/periphery organising principle of defining by difference and otherness. Instead she embraces the recognition of individuality (the fragmenting of homogeneity?):

> This is what consumers mean when they demand realism in advertising; it is projecting the woman as she wants herself to be, not as men would have her. It is through radiating power and status through a means other than a svelte figure or the vogue look. It is all about understanding her as she sees herself, rather than shackling her in the outmoded and restricted images of idealised woman . . . It means when you talk to women about [traditionally gendered pursuits] you do not assume they are different from men, you do not take men as the norm and women as the exception, [it] means you do not separate your message when talking to women . . . Individuality implies, firstly, no preconceptions about, and secondly constant monitoring of, your target market's needs. That's what women want (Adams 1991, videotape).

Conclusion

In examining core/periphery theories, and finding them more appropriate to industrial rather than post-industrial societies, this analysis rejects rigid differentiation and looks for flexibility. Postmodernism offers the more appropriate starting point for investigating globally networked societies: it represents the dynamic disjointedness which concurs with individual experience of human community. When combined with the lived experience underpinning feminist theories, postmodernism grounds its networked surface in a philosophy which is born historically from a universal experience of being peripheralised by the core. (Wherever the town centre is, at the heart of the global village, it is certain to be male.) A postmodern future based in resistance and renegotiation has the potential to move beyond rigid boundaries, enabling expression of the richness of difference.

References

Adams, Chris 1991, 'Part 2: the debate', *Women and advertising*, prods Gael Walker and Anne Ross-Smith (videotape), National Working Party on the Portrayal of Women in the Media, Office of the Status of Women, Department of the Prime Minister and Cabinet, Canberra

Ang, Ien 1991a, *Desperately seeking the audience*, Routledge, London

——1991b, 'Global media/local meaning' *Media information Australia*, no. 62, November, pp. 4–8

Angus, Ian and Shoesmith, Brian 1993, 'Dependency/space/policy: an introduction to a dialogue with Harold Innis', *Continuum: an Australian journal of culture and media studies*, vol. 6, no. 2, pp. 1–15

Bausinger, Hermann 1984, 'Media, technology and daily life' *Media, culture and society*, vol. 6, no. 4, pp. 343–51

Bell, Daniel 1973, *The coming of post-industrial society: a venture in social forecasting*, Peregrine, Harmondsworth

Brizzio, B. 1990, *Third world guide 91/92*, Institut del Tercer Mundo, Montevideo, Uruguay

Carey, James 1989, *Communication as culture: essays on media and society*, Unwin Hyman, London, pp. 201–30

Featherstone, Mike 1990, 'Global culture: an introduction', *Global culture: nationalism, globalization and modernity* ed. Mike Featherstone (*Theory, culture & society*, vol. 7, special issue) pp. 1–14

Game, Anne 1991, *Undoing the social: towards a deconstructive sociology*, Open University Press, Milton Keynes

Hamelink, Cees 1990, 'Information imbalance: core and periphery', *Questioning the media: a critical introduction*, eds J. Downing, A. Mohammadi & A. Sreberny-Mohammadi, Sage, London

Innis, Harold 1950, *Empire and communications*, Oxford University Press, Oxford

Jameson, Frederic 1984, 'Postmodernism, or the cultural logic of late capitalism', *New left review*, no. 146, pp. 53–92

Kettle, Martin and Phillips, Melanie 1993, *The Guardian 2*, tabloid section, Feb. 16, pp. 2–3

McArthur, Colin 1985, 'Scotland's Story', *Framework*, no. 26–7, pp. 64–74

McLuhan, Marshall and Fiore, Quentin 1968, *War and peace in the global village*, Bantam Books, New York

Marx, Karl 1976, *Capital*, vol. 1, Penguin, Harmondsworth

Meyrowitz, Joshua 1985, *No sense of place: the impact of electronic media on social behaviour*, Oxford University Press, New York

Oakley, Ann 1974, *Housewife*, Allen Lane, London

O'Kane, Maggie 1993, *The Guardian outlook*, Feb. 20–1, p. 21

Schiller, Herbert 1991, 'Not yet the post-imperialist era', *Critical studies in mass communication*, vol. 8, no. 1, March, pp. 13–28

Steenbergen, Pam 1993, personal communication

Wallerstein, Immanuel 1979, *The capitalist world economy*, Cambridge University Press, Cambridge

——1980, 'The future of the World-Economy', *Processes of the world system*, eds Terence Hopkins & Immanuel Wallerstein, Sage, Beverly Hills

Wark, McKenzie 1990, 'Europe's masked ball: east meets west at the wall', *New formations*, no. 12, Winter, pp. 33–42

Annotated bibliography

Ang, Ien 1991, *Desperately seeking the audience*, Routledge, London
A stimulating introduction to postmodern ideas concerning the existence of a powerful audience able to create its own meanings.

Hamelink, Cees 1990, 'Information imbalance: Core and periphery', *Questioning the media: a critical introduction*, eds J. Downing, A. Mohammadi & A. Sreberny-Mohammadi, Sage, London
Cees Hamelink analyses the gross disparity in access to information and communication technologies experienced by people of the third world, leaving the reader in no doubt that core and periphery distinctions are appropriate for comparisons between the first and third worlds.

Jameson, Frederic 1984, 'Postmodernism, or the cultural logic of late capitalism', *New left review*, no. 146, pp. 53–92
This seminal essay discusses the nature of postmodernism in an interesting and accessible form—an engaging introduction to this challenging conceptual framework.

Schiller, Herbert 1991, 'Not yet the post-imperialist era', *Critical studies in mass communication*, vol. 8, no. 1, March, pp. 13–28
Herbert Schiller has little sympathy with postmodern approaches to audience studies. In this excellent paper he argues not only that cultural imperialism exists but that it is a result of global domination by TNCs.

Wark, MacKenzie 1990, 'Europe's masked ball: east meets west at the wall', *New formations*, no. 12, Winter, pp. 33–42
A thoroughly postmodern approach to technology and culture which addresses the psychological implications of the placelessness of global communications.

▐3

VECTORY IN THE GULF: TECHNOLOGY, COMMUNICATIONS AND WAR
Michael Galvin

The history of the way wars are fought shows that changes in technology have always had a major bearing on their outcome. Likewise, the history of modern mass communications is closely tied to technological change. The contemporary newspaper, radio or television station is unimaginable without computer and communications technology—everything from satellites to telephone modems and video and audio recording equipment. The purpose of this chapter is to consider technological change in the context of both modern warfare and mass communications. In the process, four points are stressed:

- The history of mass communications technology (for example, the development of photography, then the cinema, and most recently the video camera) is much more closely tied to military needs and research than might at first appear to be the case. Indeed, it is likely that wars, more than any other cause, tend to bring together, as well as stimulate the development of, new communications technology.
- Changes in communications technology result in changes in the way human beings are able to relate with and to each other across both space and time, with different media affecting these processes of perception and communication in different ways, and with new media changing whatever was the taken-for-granted role of previously existing media. Thus, television is not just additional to radio; television changes the role and function of radio (and all other existing media, such as cinema, newspapers etc).
- The Gulf War in early 1991 is the most extreme instance to date of an increasing gap between traditional notions of 'human experience' and technological reality—a gap which, despite a massive amount of media coverage during the war, resulted in numerous major aspects of that war being barely reported at all.
- While some of the responsibility for the failure to report adequately what happened before, during, and after the war must rest with the political and military leaders of the day, the imbalance between human experience and technological reality during the Gulf War was so extreme that traditional ways of reporting the war and making sense of it for

other people (whether as television viewers, newspaper readers or, more inclusively, as citizens in those countries involved) were incapable of conveying what was taking place.

War, human experience and technological change

Nearly twenty years after the end of the First World War, the German writer, Walter Benjamin, made the following comments about its significance:

> With the (First) World War a process began to become apparent which has not halted since then. Was it not noticeable that men returned from the battlefield grown silent—not richer, but poorer in communicable experience? . . . For never has experience been contradicted more thoroughly than strategic experience by tactical warfare, economic experience by inflation, bodily experience by mechanical warfare, moral experience by those in power. A generation that had gone to school on a horse-drawn streetcar now stood under the open sky in a countryside in which nothing remained unchanged but the clouds, and beneath those clouds, in a field of force of destructive torrents and explosions, was the tiny, fragile human body (Benjamin 1973, p. 84).

Even though half a century has passed since Benjamin wrote these words, their significance and relevance has, if anything, increased. The particular issue he was dealing with in that essay concerns how human beings can still tell stories to one another, which attempt to make sense of experience when such experience appears to be so at odds with events on any easily understood human scale. 'Stories' in this context refers not just to fictions and legends but also to journalistic and historical attempts at meaning creation and the sharing of understanding.

The French writer, Paul Virilio, is among those who have drawn attention to the fundamental changes to traditional ideas of warfare which modern technology has brought about:

> If the First World War can be seen as the first mediated conflict in history, it is because rapid-firing guns largely replaced the plethora of individual weapons. Hand-to-hand fighting and physical confrontation were superseded by long-range butchery, in which the enemy was more or less invisible save for the flash and glow of his own guns. This explains the urgent need that developed for ever more accurate sighting, ever greater magnification, for filming the war and photographically reconstructing the battlefield; above all it explains the newly dominant role of aerial observation in operational planning (Virilio 1989, pp. 69–70).

This chapter is concerned with making links between technology and war on the one hand, and technology and communication (particularly mass communication) on the other. In the process, parallels between the development of military technology and the development of communications technology become more apparent.

For example, as Virilio makes clear, ways of seeing (and therefore, ways of recording and transmitting what one has seen) are as crucial to the military as they are to the media of mass communication. To see is to be able to target; it is also to be able to know whether the target has been destroyed or not. In Virilio's words, 'For men at war, the function of the weapon is the function of the eye' (Virilio 1989, p. 20). A line of sight, in other words, is a line of fire. It is a perceptual trajectory along which information flows in one direction and deadly force travels in the other. Virilio describes such a line as a vector. As we shall see later, mapping just some of the vectors at work in the Gulf War shows two important developments: firstly, the extent to which modern communications technology has transformed relations between time and space; secondly, the extent to which communications technology has contributed to the phenomenon of globalisation, not only militarily but also in the nature of 'media events'.

Since the invention of photography in the nineteenth century, recorded images have been used for the purposes of conducting war as well as for recording the conduct of the war for other purposes (informing the civilian population of what was going on and so on). However, the nature of visual perception achieved a new and deadly synthesis during the Gulf War. Here is how one military analyst has described what was so different about the Gulf War:

> First, the coalition strategists became arguably the first wartime generation of strategists to have unbroken optical contact and other sensory contact with their enemy. Second, given the sophistication and definition of images provided by reconnaissance satellites, they had the confidence that they could 'believe their eyes'. Third, in view of the fact that virtually nothing could escape being sensed, what could be sensed could be targeted; and what could be targeted could be bombed, rocketed, strafed—in a word, killed. And fourth, since intelligence was available without interruption and in 'real time', a single composite was formed consisting of sensing, detecting, targeting and killing. Fused were the human eye, the satellite eye, the target and the selected weapon: it was the nearest example history has seen of the ultimate hand-eye co-ordination (McKinley 1991, pp. 186–7).

The changes to our usual ideas of what 'vision' entails as described here by McKinley are so fundamental that their significance is easily overlooked. For example, we are so used to associating night and darkness with concealment that it is difficult to imagine what it must have been like to be able to operate with full visibility 24 hours a day, as the allied forces were able to do with their sophisticated thermal cameras and electro-optical lighting techniques. Nevertheless, the consequences for the Iraqi soldiers were devastating.

To take only one example, the US Army's Apache attack helicopter proved to be one of the most deadly machines in the war. The Apache's pilots were guided by an infra-red optical system that turned blackness into a bright phosphorescent daylight—although only on their on-board video monitors—in which, in the words of one reporter, 'you can all but read the expressions of shock on the faces of the Iraqi soldiers as they are hit by cannon rounds and rockets' (*Guardian Weekly* 1991a, p. 10). (In some cases, these screens were located in the visors of the crews' helmets.) Video has become both the medium of 'real-time' weaponry as well as the medium of recorded information and feedback. By means of video cameras built into the guns on board, the pilots take aim by looking at a video screen—a screen on which distant Iraqi soldiers, for whom their environment is still pitch black, appear as big and clearly delineated as football players on a home television screen. Because the helicopter is hovering less than twenty metres above the ground somewhere in the distance, the Iraqis have nowhere to run to because they have no idea what to hide from. They are blown apart on the video screen in real time; but the video system has also recorded their deaths in graphic close-up—for use later in assessing the effectiveness of weapons systems and to aid in training sessions.

It is easy to describe the vectors at work in the attack helicopter example. A communication loop exists between the gunner via his video screen, connected to his gun-sights/camera, and his human target in the desert some distance away (even though the Iraqi victim is not likely to realise that he is caught in this communication/information loop until it is too late). Of course, there is no reason why the vector should stop there. Connection in real time with a command base would be possible, although it is not clear whether such connections were being made on any large scale during the Gulf War. And, if connection with a command base is possible, then it would be a relatively simple step for that vector to extend to television sets 'live' in living rooms around the world.

Of course, the word 'live' disguises the grim reality of what is being described here: the technical means to enable people around the world to watch other people being killed in real time and in visual detail by means of the same technology which makes the killing possible in the first place. To watch such events 'live' in another country, or even to be the helicopter gunner who has an image of Iraqi soldiers as if they are close by, when in fact they might be kilometres from him and completely invisible without the technical means at his disposal to 'illuminate' and 'magnify' them and turn them into clearly delineated video images, is to change space relations as they are normally experienced on a human scale. (Because so much of what happened in the Gulf War took place by means of video screens and monitors, it is easy to see why that conflict quickly acquired the title of the 'Nintendo War').

The other important feature of this weapons/communications system is that it makes it impossible to distinguish a training simulation from the 'real thing'. In terms of the sensory experience of the crew, simulation and execution have become indistinguishable—with both the practice and the action bearing greater resemblance to a computer game than to how combat took place or was prepared for even twenty years ago.

Likewise, to be able to replay such images—for whatever purposes—and to be able to manipulate them (fast forward, freeze, slow motion, reverse etc.) is to change the experience of time in any conventionally understood sense.

Within communication studies, it has long been acknowledged that successive changes in the media of communication have consequent effects on space/time relations. Such a relationship has been most systematically explored by the Canadian theorist, Harold Innis, in books which are now classic studies in the history of communications (even if his arguments, which are now nearly half a century old, are not to be taken wholly uncritically) such as *The bias of communication* (1951) and *Empire and communications* (1950). In Innis' words:

> A medium of communication has an important influence on the dissemination of knowledge over space and time . . .
>
> According to its characteristics it may be better suited to the dissemination of knowledge over time than over space, particularly if the medium is heavy and durable and not suited to transportation, or to the dissemination of knowledge over space than over time, particularly if the medium is light and easily transported. The relative emphasis on time or space will imply a bias of significance to the culture in which it is embedded (Innis 1951, p. 33).

Media determinism?

As this excerpt suggests, Innis divided communication and social control into two major types: those achieved through space-binding media and those achieved through time-binding media. Because they are easy to transport, space-binding media, such as print and electronic communication, are connected with expansion and control over territory and favour the establishment of commercialism, colonisation and empires. Because they are more difficult to transport, time-binding media such as the manuscript and human speech, favour the cultivation of memory, a historical sense, relatively small communities, and traditional forms of authority.

Whereas print solved the problem of producing standardised communications rapidly and in sufficient quantities to administer large areas, the development of electronic communication—beginning with telegraphy and moving from radio to television—solved simultaneously the problems of rapid production and distribution. These technologies not only eclipsed space but transformed time, eventually obliterating memory and reducing message duration to the hour, minute, second, and microsecond (see Carey 1989, pp. 142–72).

Both Innis, and his more famous colleague, Marshall McLuhan, have been criticised because of the technological and media determinism which seems to underlie propositions such as these. A technological determinist could be defined as someone who takes the position that devices like printing technology, telephones and computers are 'humanity's prime movers', that, in Virilio's example (below), we do not just drive cars, we are driven by them! In other words, in the final analysis, structures of consciousness parallel the technologically determined structures of communication.

It is perhaps impossible to ever deal adequately with the rights and wrongs of such a position in any absolute sense, although many would wish to insist upon the overriding importance of the social and political context in which decisions about technology are made, given that technology involves decisions about how to use resources in certain ways but not others. However, it is also true, in regard to Innis' work and the technological features of the Gulf War being commented upon in this chapter, that extreme changes in space/time relations were apparent. These would include:

- the positioning of video cameras on the noses of 'smart' missiles capable of relaying images up to the very moment of impact, and providing a viewing position/space absolutely impossible for a human being to occupy.
- the use of the same pictures at the same time around the world because of the global nature of the communications systems that enabled events to happen (see below) and the systems that were in place to broadcast coverage of such events around the world.

Globalisation

Analysis of how social life is ordered across time and space leads to another concept relevant to the relationship between technology and war: globalisation. The sociologist, Anthony Giddens, has described globalisation as 'the intensification of world-wide social relations which link distant localities in such a way that local happenings are shaped by events occurring many miles away and vice versa' (Giddens 1991, p. 64).

Giddens illustrates the phenomenon of globalisation by pointing out how the increasing prosperity of an urban area in a city such as Singapore might be causally related, via a complicated network of global economic ties, to the impoverishment of a neighbourhood in Pittsburgh (or Melbourne, for that matter) whose local products are uncompetitive in world markets.

Of course, globalisation applies just as much, if not more so, in the technological dimensions of what is focused on throughout this chapter: military technology and mass communications technology. Regarding the latter, the world-wide prominence and reach Cable News Network (CNN) enjoyed during the Gulf War is well known; even though that particular instance is only an obvious example of general trends in all media industries whether oriented primarily to information or to entertainment.

Putting the Gulf War into this globalisation context, it is clear that the vectors of military information/action crossed national borders and even continents as a matter of course, and as McKenzie Wark has pointed out, made it quite difficult to pinpoint the nature of 'events' in that war:

> Did the Gulf War take place in Kuwait, Baghdad or Washington? Was the site the Middle East or the whole globe? This is a particularly vexing point. If Iraqi commanders order a SCUD missile launch via radio-telephone from Baghdad, the signal may be detected by orbiting US satellites. Another satellite detects the launch using infra-red sensors. Information from both will be down-linked at Nurrungar in South Australia. From there it will be relayed to the Pentagon, then again to US command HQ in Saudi Arabia and to Patriot missile bases in Saudi Arabia and Israel (Wark 1991, pp. 6–7).

Speed and politics

The logistics of perception which affect our relations to both the spatial and the temporal dimensions of our existence are constantly being changed by technical innovations, and there is nothing new about this process. For example, many people cannot see without glasses or hear without hearing aids. (Perhaps the day is not far off when many of us cannot count without a calculator or think without a computer?) These prosthetic devices bring about a significant change in such people's relationships with their environment.

However, it has also been argued that this process of change has been so greatly speeded up in recent times that mechanical and then electronic technology has introduced a fundamental and crucial division between what Paul Virilio has called 'metabolic speed' and 'technological speed':

> There is a struggle, which I tried to bring to light, between metabolic speed, the speed of the living, and technological speed, the speed of death which already exists in cars, telephones, the media, missiles. There is also a couple formed by the metabolic speed of the living and the technological speed of the

deterrence. Politics should try to analyse this interface (Virilio 1983, pp. 140–1).

Two of Virilio's books (*Pure war* 1983, and *Speed and politics* 1986) are devoted to his concern with the significance of speed as a concept of extreme importance, not only in warfare, where it is obviously so, but also in political and economic life. According to Virilio, speed must be politicised, whether it be metabolic speed or technological speed 'because we are both: we are moved, and we move. To drive is also to be driven' (1983, p. 30).

Returning this discussion to the Gulf War, extreme speed characterised the gathering of military information, the manner of attack, and the manner of its communication whether within the decision-making loop or what was permitted to be shown on broadcast television, with these three 'activities' becoming integrated and functioning in 'real time' more than ever before. Even so it is important not to assume that the communications technology which enabled this convergence to take place actually resulted in television viewers of CNN—or the other networks—experiencing the war just as the allied decision-makers were experiencing it. As already discussed, the video coverage obtained by the attack helicopters was extremely graphic, but almost none of this was ever shown on broadcast television.

Nevertheless, as we have seen, all three interdependent systems depended on 'high-tech' hardware and software operating on a global scale. It is time, therefore to turn to how this war was actually reported, and to examine whether links can be made with the nature of this reporting and such a 'politics of speed'.

Telling the story of the Gulf War

During the Gulf war in early 1991, many television channels around the world broadcast war-related programming 24 hours a day. The most famous of these broadcasting networks was undoubtedly CNN, an American cable television news network which benefited, firstly, because this type of event was one which they were specifically set up for (news consisting of 'live crosses' to on-the-spot correspondents, experts of various kinds, and political leaders being their staple programming, not something which gained extensive airtime only because of the exceptional and newsworthy nature of what was taking place), and, secondly, because they managed to keep a western reporter in Baghdad after nearly every other journalist had been forced to leave Iraq.

So pervasive was the television coverage that Marshall McLuhan's metaphor of the 'global village' was resurrected to describe this, 'the most televised war in history'. Jonathan Holmes, a television producer with the ABC, describes the experience in retrospect thus:

> What the Americans—indeed, what all of us—specialised in was talk. During the first few days of the war, especially, the world-wide satellite links allowed whole populations across the planet to plug in for hours at a time to the live feeds from the great American newsrooms. Professionals elsewhere watched, awed, as their superbly smooth anchors—Peter Jennings and Ted Koppel, Tom Brakaw, Dan Rather—knowledgeable, unflustered, almost indecently articulate, ad-libbed their way through interview after interview around the world. 'I'm sorry, Senator, we'll have to leave it there, we're crossing now to our reporter Joe Shepton in Jerusalem.' And Joe would be there, telling us live about the latest Scud attack and what he could see from his window. Across, live, for reaction in the streets of Cairo, or the Jewish quarter of New York. A missile expert in Santa Monica followed by a national security analyst in Washington, back to the patient senator for a quick reaction before crossing to catch an impromptu press conference being given by John Major on the doorstep of No. 10 (Holmes 1991, p. 198).

Largely as a result of its coverage of the Gulf War, CNN received unprecedented ratings and publicity, culminating in its founder Ted Turner being proclaimed by *Time Magazine* (1992, pp. 6–9) as its 1991 Man of the Year (and described in its covering story as 'Prince of the global village').

Given the sophisticated military and mass communications technology made startlingly apparent during the Gulf War (and hopefully enough has been said throughout this chapter to indicate that it is more useful to see these two forms of technology as fundamentally linked rather than as separate and discrete domains), it is disturbing that some of the most important issues and events in that war went virtually unnoticed and that media distortions of historical reality occurred which would normally have been considered blatant propaganda.

For example, there was little or no media coverage of any historical background to the dispute between Iraq and Kuwait that led eventually to war, even though the border between the countries has been bitterly contested since the British High Commissioner in 1923 placed a large notice board at the southerly edge of some date palms with the words 'Iraq–Kuwait Boundary' inscribed thereon. The potential for conflict inherent in the British Commissioner's opportunistic carving up of the desert can be gauged from the fact that Iraq was given a narrow coastline of some 50 kilometres, with its outlet to the gulf almost blocked by two adjoining Kuwaiti islands, whereas Kuwait, little more than a city-state, and with a population less than one eighth Iraq's, ended up with a coastline of some 500 kilometres (Draper 1992a, pp. 46–7).

One would have looked in vain in most of the western media during the crisis to have found any sense that the conflict was any more complicated than a straightforward case of a belligerent bully picking on a defenceless and innocent neighbour, and thereby deserving to be thrown out violently regardless of how many lives might be lost in the process.

Despite the high-tech features of the war, much of its coverage in the media was as distorted and propagandist as any previous conflict. For example, it has been shown that, in one single week during the conflict, the British press used the following terms to describe allied troops and Iraqi troops:

> Allied troops: boys, lads, professionals, lion-hearts, cautious, confident, heros, dare-devils, young knights of the skies, loyal, desert rats, resolute, brave.
>
> Iraqi troops: troops, hordes, brain-washed, paper tigers, cowardly, desperate, cornered, cannon fodder, bastards of Baghdad, blindly obedient, mad dogs, ruthless, fanatical (*Sunday Age*, February 10 1991, p. 8).

However, such blatant propaganda and disregard for the full facts has probably been true of many wars and most participants for as long as the power of the media to influence events has been recognised. What was not so predictable, and now appears much more shocking, was the inability of the media, despite huge resources, to be able to give any sense at all of what happened to the Iraqis during the conflict.

To understand why the media failed in this basic duty, it is necessary to return to the ideas of Walter Benjamin with which this chapter began. Benjamin was describing what to him was the most significant fact of the modern age: a fundamental gap between a technological scale of events and human experience unmediated by such technologically advanced systems. Benjamin's interest was in what this gap does to the capacity of human beings to tell one another meaningful stories, stories in which human experience was recognisable even if not necessarily moral. Benjamin summed the dilemma up in a famous couple of sentences:

> More and more often there is embarrassment all round when the wish to hear a story is expressed. It is as if something that seemed inalienable to us, the securest among our possessions, were taken from us: the ability to exchange experiences (Benjamin 1973, p. 83).

If we combine this idea of Benjamin's with the importance Innis attaches to media of communication in changing our sense of location in both time and space, and with Virilio's additional emphasis on the significance of speed itself in transforming our 'logistics of perception', then another way of explaining the gaps and distortions which characterised the media coverage of the Gulf War might be possible, apart from the obvious facts of media manipulation which also occurred.

Is it possible that some at least of what happened in the Gulf War was so unprecedented, so radically different from ways of thinking about military experience in the past, that such 'experience' was, for a time at least, incommunicable and unnarratable? That the ability to exchange experiences had been lost, at least for that time while there was still much public interest in the war?

This alarming scenario is supported by two aspects of the reporting of the war: the reporting of casualties, and the reporting of the final hours of battle.

As the ABC producer noted above, television coverage of the war consisted mainly of talk supplemented by video footage of missiles heading for targets, with vision lost at the moment of impact and subsequent destruction. What television did not show was death, individuals suffering, blood, or the mutilation of human bodies. Yet it has been estimated that, while fewer than 200 allied personnel were killed, between 100 000 and 200 000 Iraqis (mainly conscripted soldiers) were killed—a ratio of Iraqi to allied losses of a thousand to one (Draper 1992b, pp. 38–45). Words such as 'war' or 'battle' scarcely do justice to the causing of death on such a massively unbalanced scale. A more usual word, based on our historical experience, would be 'massacre', yet one would look in vain to find much reporting of the war in such terms.

Yet 'massacre' seems the only appropriate word to describe the slaughter which took place in the final hours of the war, as the Iraqis, taking an unknown number of Kuwaiti hostages with them, fled Kuwait in whatever vehicles they could (buses, taxis, stolen cars and trucks, as well as armoured vehicles and personnel carriers), only to be trapped on a bombed freeway in the middle of the desert, and destroyed by the helicopter gunships and planes which have already been discussed. The number killed during those hours has never been disclosed, but could have been in the tens of thousands. While there were some accounts of this totally unnecessary tragedy (given that Kuwait had by then been abandoned by the Iraqis) published in subsequent months (for example, 'Trapped in the killing ground at Mutlaa', *Guardian Weekly*, 1991b, p. 17), the lack of reporting in the Australian and American mainstream media was conspicuous.

What was reported around the world were photographs which showed burnt-out vehicles clogging the highway. In *The Australian*, for example, such a picture was captioned 'An Iraqi convoy, laden with booty, destroyed by allied bombing on the road to Basra' and was placed on the front page under the main headline, 'Iraqi torturers escape allied net' (*Weekend Australian*, March 2–3, 1991). Some months after the war, when News Limited produced a special 32-page colour supplement called *Gulf War diaries* (News Ltd 1991), the only reference to this slaughter was a similar photograph of wrecked vehicles on a desert highway. Hard as it might be to believe, the events on that highway—which one American pilot described as 'like shooting fish in a barrel' (*Guardian Weekly*, March 17 1991, p. 17)—did not receive one sentence of reporting in the whole 32 pages.

How to explain such omissions, such irresponsible reporting, such lack of respect for, or even interest in, the Iraqi dead? Media manipulation and propaganda can explain some of what happened, but the argument being advanced in this chapter goes one step further: that communications technology converged with military technology to such an extent that, not only did the nature of battle change, but it changed in such a way that it was so outside the experience of most people (including reporters) that what really happened was, at least temporarily, literally unreportable. Further, that such unreportability was made more difficult to detect precisely because the communications systems were so advanced, and so obviously present, that an absence of significant information about what was going on was made to seem that much more implausible. It is a more difficult task to convince someone who has watched weeks of 'authoritative' television coverage of an event that they do not know enough, than it would be to convince someone who has watched no coverage at all.

Another way of looking at this point is with reference to the concept of discourse, a term frequently used in communication studies, and which we might define as systematic frameworks of both words and ideas which are drawn upon and used to encode, or make narratives about, some aspect of reality. It is not difficult to see two opposing (but in this case largely irrelevant) discourses at work in the words used to describe Iraqi soldiers compared to the words used to describe the allied soldiers. War has a number of common discourses, as this chapter has discussed. Technology has also. The point being made here is that, even though these two discourses have often been intertwined, such an intertwining this time round resulted in very partial accounts of the Gulf War, and not just for propagandist reasons. The discourse of technology was too removed from military discourses which enable wars to be justified, or even described, in human terms.

The future?

Manuel De Landa, at the start of his book, *War in the age of intelligent machines* (1991), written after the Gulf War had introduced television viewers around the world to the notion of 'smart bombs' and 'missile-cam', introduces the reader to the Prowler.

The Prowler, as described by De Landa, is a small terrestrial armed vehicle, equipped with a primitive form of 'machine vision' (the capability to analyse the contents of a video frame) that allows it to manoeuvre around a battlefield and identify enemies. The Prowler is now at the working prototype stage. According to De Landa, the Prowler still has difficulty making sharp turns, travelling over rough terrain, or even distinguishing friend from foe. De Landa continues:

> For these reasons it has been deployed only for very simple tasks, such as patrolling a military installation along a predefined path. We do not know whether the Prowler has ever opened fire on an intruder without human supervision . . . (De Landa 1991, p. 1).

The Prowler is really little more than a ground-hugging version of the 'smart bombs' used in the Gulf War. But what it throws into sharp relief is the technological capacity to use machines not only to kill people, or to advise on the killing of people (by providing relevant information), but to have the machine make such a decision, as well as have the ability to carry it out.

The moral questions raised by this ability of technology to 'exteriorise' human decision-making processes are indeed profound. What is perhaps pointed to is a widespread confusion of boundaries between humanity and electronic technology, a confusion which makes such events as the Gulf War exceedingly difficult to comprehend within the types of narrative which have been used until now to make sense of wars.

Conclusion

Narratives need some form of agency. We have become so used to agency equating with human characters (heroes, villains, innocent bystanders and so on) that it is difficult to absorb the inadequacy of such a person-centred perspective to such realities as the Gulf War.

However, it is also true that human beings will continue to try to make sense of their environment and their experience, even if the nature of both can only be guessed at. A feature of many popular recent films has been the character who has been part-human, part-machine: a cyborg, in other words (cyborg standing for the combination of cybernetic and organism). Such films would include *Bladerunner*, the *Terminator* and *Robocop* series, and *Total recall*, to name some of the most commercially successful.

Rather than consigning such films to the category of escapist or teenage male fantasy (and seen as harmless or harmful depending on what discourse is brought to bear on them), perhaps they are consciously or unconsciously making attempts to explain a world in which a high-tech war such as the Gulf War is not only possible but now a sad reality. One thing, however, is certain. There is a great need in our culture for stories which do justice to the issues raised by the use of intelligent machines in warfare—not just how to think about such matters morally or ethically, but how to think about them at all!

References

Benjamin, W. 1973, *Illuminations*, Fontana, London, pp. 83–110

Carey, J. 1989, *Communication and culture*, Unwin Hyman, London

De Landa, M. 1991, *War in the age of intelligent machines*, Swerve Editions, New York

Draper, T. 1992a, 'The Gulf War reconsidered', *The New York review of books*, January 16, pp. 46–52

——1992b, 'The true history of the Gulf War', *The New York review of books*, January 30, pp. 38–45

Giddens, A. 1991, *The consequences of modernity*, Polity Press, Cambridge

Guardian Weekly 1991a, March 3, p. 10

——1991b, March 17, p. 17

Holmes, J. 1991, 'The media war', *43 days: the Gulf War*, eds I. Bickerton and M. Pearson, Australian Broadcasting Corporation, Sydney

Innis, H. 1950, *Empire and communications*, University of Toronto Press, Toronto

——1951, *The bias of communication*, University of Toronto Press, Toronto

McKinley, M. 1991, 'The battle', *43 days: the Gulf War*, eds I. Bickerton and M. Pearson, Australian Broadcasting Corporation, Sydney

News Ltd 1991, *Gulf War diaries*, Nationwide news, Canberra, March

Sunday Age 1991, February 10, p. 8

Time Magazine 1992, 'Prince of the global village', January 6, pp. 6–9

Virilio, P. 1983, *Pure war*, Semiotext(e), New York

——1986, *Speed and politics*, Semiotext(e), New York

——1989, *War and cinema: the logistics of perception*, Verso, London

Wark, M. 1991, 'News bites: war TV in the Gulf', *Meanjin*, vol. 50, no. 2, pp. 5–17

Weekend Australian 1991, March 2–3

Annotated bibliography

Benjamin, W. 1973, *Illuminations*, Fontana, London

Benjamin provides numerous perceptive insights into the relations between culture, media and technology.

De Landa, M. 1991, *War in the age of intelligent machines*, Swerve Editions, New York

This book shows that the recent emergence of intelligent and autonomous bombs and missiles is part of a larger transfer of cognitive structures from humans to machines.

Gibson, W. 1988, *Burning chrome*, Grafton Books, London

A stimulating introduction to the genre of cyberpunk fiction. Gibson's trilogy: *Neuromancer, Count zero,* and *Mona Lisa overdrive* (all Grafton) is also worth reading.

Innis, H. 1951, *The bias of communication,* University of Toronto Press, Toronto
Innis' now classic elaboration of his argument that communication media tend towards a bias either of a 'time-binding' or of a 'space-binding' kind.

Robins, K. and Levidow, L. 1991, 'The eye of the storm', *Screen,* vol. 32, no. 3, pp. 324–8
This lucid and thoughtful essay tries to make sense of the Gulf War, and the ends to which communications technology has apparently been put.

Springer, C. 1991, 'The pleasure of the interface', *Screen,* vol. 32, no. 3, pp. 303–23
An excellent essay analysing the currency, in popular culture and the scientific community, of a discourse describing the union of humans and electronic technology.

Virilio, P. 1989, *War and cinema,* Verso, London
An eclectic, sustained argument about interrelationships between modern warfare and modern mass communications.

Wark, M. 1991, 'News bites: war TV in the Gulf', *Meanjin,* vol. 50, no. 2, pp. 5–18
An excellent application of Virilio's ideas to issues raised by the Gulf War, with particular reference to the role of television.

A SUSTAINABLE FUTURE: LIVING WITH TECHNOLOGY

Adrianne Kinnear

In 1987, a report commissioned by the United Nations (WCED 1987), *Our common future*, placed the issue of environmental sustainability firmly on the political agenda. Sustainability was set to become the buzzword of the 1990s. But what exactly do we mean by sustainability and how is it to be achieved? Is it compatible with economic growth and development? What is the relationship, if any, between our use of technology and environmental sustainability?

The concept of sustainability is a little like beauty—it's very much in the eye of the beholder and there is a confusing societal mix of perspectives of sustainability. It's useful to have some kind of classification scheme.

We can place the major perspectives of sustainability on a continuum from biocentric (the environment at all costs) at one end, to technocentric (technology at all costs) at the other (Folke & Kaberger 1991). Four main perspectives encompass most current viewpoints:

- The *extreme greenie* perspective is biocentric and views the human organism as having no more worth than other biological organisms. All species have an equal right to existence and any directions towards sustainability or the use of technology in sustainable systems must recognise the rights of all species to persist. Preservation of the environment subsumes all use of technology and sustainability. Some proponents recommend a return to the solar-powered and environmentally compatible behaviours of earlier human societies.
- The *technofix* perspective is the extreme alternative. It views the environment as a set of mineable resources to be used for the economic benefit of humans. Growth is limited only by our creativity for technological innovation. The solutions to environmental degradation and the finite nature of resources will depend on technological innovations to handle waste and emissions and to identify and to continually substitute new types of resources, goods and services for old ones.
- The *resource development* perspective is the one which most industrialised nations have adopted. Like the technofix view, it is anthropocentric. Its primary focus is the identification and development of resources for continued economic growth. However, it does recognise a secondary

value for the 'environment' in that development should proceed so as to minimise environmental disruption and biodiversity reduction. It incorporates an environmental protection theme but states that economic growth must be a given in any sustainable future, and while it should incorporate at least some environmental concerns it should not be retarded by them.

• The *resource management for development* perspective is one which carries much political currency so far in this decade because it reflects the perspective of *Our common future*. While still anthropocentric in focus, it recognises that environmental protection is essential for sustainability (since the environment provides basic life-support systems), and that maintenance and protection of ecosystem processes, including biodiversity, must be a primary consideration in any technological development for economic growth. It argues for a greater incorporation of ecological principles into economic systems than do the other anthropocentric perspectives.

Ecosystems as model sustainable systems

We are surrounded by systems which have evolved to use energy and resources in a sustainable way. They are the natural communities of animals and plants which biologists call *ecosystems*. We can imagine ecosystems as the living organisms of an area interacting with the surrounding non-living components (air, water, soil). This interaction is two-way. The structure of the living communities is determined to some extent by the air, water and soil, and the physical environment is, in turn, modified by the actions of its inhabitants. Ecosystems can vary enormously in size, complexity and biology. A compost heap, a local wetland, a jarrah forest, the biosphere, can all be described as ecosystems. What they have in common is an ability to sustain a continuously evolving variety of life forms—the biosphere's biodiversity. They do this by:

• processing solar energy into a biologically useful form of energy currency (ecosystem food); and
• mobilising and exchanging a small portion of the earth's finite matter resources (ecosystem nutrients).

Ecosystems rely on the continuous flow of energy through them via a unidirectional set of biological pathways called food chains. It begins when a tiny fraction of solar energy is transformed by photosynthesising green plants into chemical energy in plant tissue. This plant tissue becomes the energy source for the plant-eaters of the system who transform part of it into chemical energy contained in their own body tissue. This animal tissue in turn becomes the energy source for the animal-eaters. Thus energy (disguised as food) moves through the ecosystem along the who-eats-whom highways above and below the ground.

The organisms are the energy transformers of the ecosystem, providing the interconnecting links between and within all the parts. Thus vegetation may be eaten by a cricket, which is then consumed by a bandicoot which is taken by a bird of prey. This energy transformation process continues below ground where the decay organisms feed on excreta, and dead animal and plant matter, obtaining energy for their own biological processes. The decay process is complete when all the useful biological energy has been utilised, the conversion to heat energy is over, and only the 'ashes'—the inorganic building blocks of organisms—remain.

But even with natural systems, all is not as perfect as it seems. There are natural laws governing the energy and matter relationships of all the biosphere's operations, the laws of thermodynamics. These laws tell us that, while we can transform matter or energy into different forms, we can neither create it nor destroy it. Furthermore, whenever such a transformation takes place, some energy is always converted into heat and dissipated to the environment.

Ecosystems are notoriously inefficient at transforming energy from one organism to another. Take yourself as an example. For every 100 g of food you eat, approximately ten g of that meal is transformed into actual body tissue. The remainder (representing the energy contained in the other 90 g) is channelled off through a variety of other kinds of chemical transformations to provide energy for movement, for body repair and maintenance and for carrying out all the other life-support processes (digestion, excretion, circulation). Ultimately, this remainder is converted into low-temperature heat energy and dispersed into the environment.

Moving from the individual level to the ecosystem level, this means that, of the total energy transformed into plant or animal tissue by the ecosystem, only 10 per cent of it can be made available as food at the next feeding level. This is the energy cost for maintaining the complex biodiversity of the biosphere. Thus, as one proceeds along a food chain, the fraction of available energy transformed into biologically useful energy decreases rapidly. Simultaneously, the amount of energy transformed into low-temperature heat energy increases. The energy flow is one-way because the low-temperature heat energy cannot be re-transformed (recycled) into higher grade energy without additional energy inputs. Because of this, ecosystems rely on a perpetual supply of solar energy.

Energy alone will not support life in ecosystems, however. Its role is to enable the organisation and rearrangement of a select group of essential chemicals into animal and plant tissue. These chemicals (nitrogen, phosphorous, oxygen and the other essential components we know as nutrients) are a very tiny subset of the biosphere's finite material resources and are the

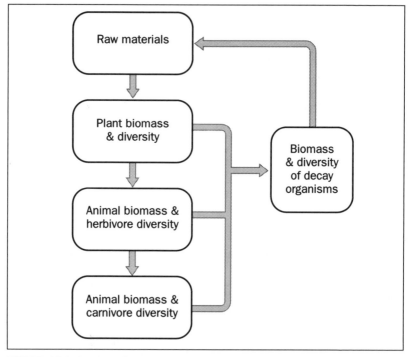

FIGURE 14.1 A generalised ecosystem
A representation of an ecosystem as a linear series of cyclical
biological matter transformations powered by solar energy.

matter-base of ecosystems. While it is easy to see that chemical and nutrient
resources will travel through the ecosystem along the food chains, paralleling
energy flow, there are some important differences.

Firstly, the primary source of these resources is the biosphere itself—the
water, soil and rocks. Organisms mobilise these resources from soil and water
and they enter the biological pathways through plant roots. Secondly, unlike
the perpetual supply of solar energy, the biosphere's supply of these essential
resources is finite. Matter must be recycled over and over if life is to be
sustained.

The pathways of chemicals and nutrients through and between ecosystems are true cycles. The biological links ensuring the recycling process are
the decay organisms which break down dead plant and animal matter into
components which become available for plant uptake and re-use by the
system. Without these decay pathways the ecosystem cycle would cease.
Biological resources would remain locked away from further use in dead

plant and animal bodies, and environmental sustainability would grind to a halt.

Sustainable ecosystems perform a variety of indispensable environmental services (Folke 1991):

- maintenance of the chemical quality of the soil, air and water;
- ensuring the thermal integrity of the biosphere and hence the nature of the climate;
- ensuring the appropriate distribution for re-use of water and essential resources;
- maintenance of genetic biodiversity;
- ensuring the breakdown and assimilation of wastes.

These environmental services are jeopardised by our current use of non-sustainable technology.

Technosystems as non-sustainable systems: the agrosystem example

The history of technology, from an environmental viewpoint, is really a story about changes in the flows of energy and matter which we humans have superimposed on the natural ecosystem flows. To appreciate how our use of technology has altered the pathways of energy and resources in the biosphere consider the evolution of the agrosystem—a technological system dedicated to food production. Humans have 'progressed' from food-getting ecosystems to a complex technosystem relying on substantial flows of supplementary energy (such as motor fuel and electricity) to power increased numbers of energy-using steps between soil and table. Material now flows one way; it is not recycled. 'Waste' is dissipated rather than re-used. The progression from early, sustainable agrosystems to the current unsustainable situation is outlined below.

Hunter gatherer agrosystems

The nomadic Kung bushmen of the Kalahari desert are thought to be characteristic of early human societies (Sabath & Quinell 1981, pp. 165–73). They neither grow crops nor herd cattle but are reliant on the native plants and animals of the ecosystem, spending much of their time gathering natural vegetable resources and hunting wild game as they require it. The productivity of this system is dependent on the amount of solar energy flowing through it and also requires human energy inputs. There is limited use of fire energy for cooking. These solar-powered hunting and food-gathering techniques are believed to have little impact on the ecosystem.

Horticultural agrosystems

The Tsembaga of New Guinea use a tiny fraction of the rain forest biomass for food (Rappoport 1971, p. 120). The process of clearing and burning small forest areas for simple cultivation of native plants provides almost all their dietary needs. Their solar-powered horticultural techniques, with human energy inputs of clearing, seeding, weeding, harvesting etc. and fire energy, mimic ecosystem function. Each cleared area is abandoned after a period of time and the forest regenerates. Native animals and plants are also hunted and gathered. Feral pig husbandry provides additional protein and the pigs assist in the consumption of waste materials.

Early agricultural agrosystems

The small rice plot of monsoon climates was, and still remains, one of the most successful systems for supporting large populations on solar energy. Moving away from reliance on the native plants and animals, humans developed seed crops of cultivated grains such as rice. Native plants remain a food source for domesticated breeds of animals such as cows or water buffalo. Such animals subsidise human energy inputs for soil preparation, crop cultivation and the transport of food produced to town dwellers. Animal and plant wastes replace soil nutrients which are removed with cropping. This agrosystem is 'efficient' enough to permit the production of a surplus which can sustain a town-based population.

Modern agrosystems

The greatest changes to our food-getting systems came with the development of technology enabling us to subsidise solar, human and animal energy with fossil fuels. In the modern agrosystem characteristic of industrial nations large amounts of fossil fuels flow through the system, fuelling every step in the chain. Technologies for the processing and packaging of food have increased the numbers of steps (transformations) in the path from farm to table. Crops are cultivated, animals farmed, produce is harvested or slaughtered, preserved, packaged, transported, stored, cooked and consumed. Large and diverse amounts of matter source the system (fertilisers, pesticides, packaging, transport and retail infrastructures) with little being recycled or re-used. For each unit of energy reaching our tables, we use more than five units of energy to get it there. Much of this energy comes from a source which is finite and non-renewable—oil (Gifford 1976, pp. 412–17).

There is no doubt that technological advances in food growing, gathering and processing have provided an expanding human population with ever-increasing amounts and varieties of food. It is nonsensical to suggest that we should turn the clock back and reverse the technological gains. But

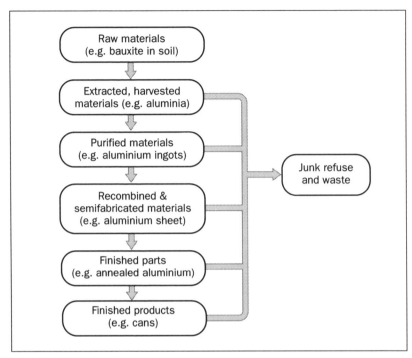

FIGURE 14.2 A generalised technosystem
A representation of a technosystem as a linear series of one-way technological matter transformations, powered mainly by fossil fuel energy. At the end of the sequence, matter is not recycled but builds up in the environment as waste (modified from Ayres 1989).

modern agrosystems are degrading the very environmental resources on which the system depends. Over the last two decades 480 billion tonnes of topsoil have been eroded from the world's farmlands, and deserts have claimed some 120 million hectares (Brown 1991). Soil erosion and desertification, and disruption of the natural matter cycles (salinity, acidification, chemical pollution of both soil and water resources), are all consequences of our current technology use. This kind of agrosystem is environmentally unsustainable (Barr & Cary 1992).

A sustainable agrosystem is likely to be one which mimics ecosystem function by reducing dependency on non-renewable energy. It will replace synthetic chemical use by natural materials which are part of the ecosystem cycle and which maintain biodiversity. Soil will be used as a valuable renewable resource rather than as a store of nutrients to be mined.

The trend towards increased flows of energy and materials with consequent environmental damage that we see in our food-getting systems is typical for most of our technosystems. The generalised technosystem, like ecosystems, is a sequence of linear transformations of energy and resources. Like ecosystems, the operations of technosystems are constrained by the laws of thermodynamics. With each transformation, heat energy is produced. As a result of our prodigious energy consumption, we pump 6 million tonnes of carbon into the atmosphere each year, and the amount of carbon dioxide is rising 0.4 per cent per year (Brown 1991). Cars contribute in large measure to this pollutant and currently we add nineteen million cars to the global fleet annually. Perhaps it is no wonder that we have just lived through the warmest decade on record—*an inevitable consequence of increased energy flow through technosystems is increased amounts of heat dissipated to the environment, biospheric warming and climate change.*

While matter may be altered as it passes through the system, it can never be destroyed. Technosystems result in products for 'consumption', but we know that matter cannot be 'consumed'. Only the utility of the product is consumed, after which we discard the product itself. The matter embodied in that product continues to exist—as waste which, if not recycled back into the system, is dissipated into, and required to be assimilated by, the environment! Municipal solid waste continues to increase 2–3 per cent annually, while mining waste produces six times the amount of municipal waste—over a billion tonnes per year (Brown 1991). *An inevitable consequence of increased one-way flows of matter through technosystems is increased waste and disruption of the biogeochemical cycles.*

In our efforts to source the enormous energy and material flows through technosystems, we are continually drawing on the finite resources of the environment (coal, oil, mineral and ore bodies). *An inevitable consequence of increased one-way flows of matter from finite resources is environmental resource depletion.*

In our efforts to transform large amounts of natural resources into providing goods and services, we utilise potentially renewable common resources (soil, air, oceans and lakes, biodiversity) in a way which renders them non-renewable (by pollution, erosion, extinction). Air and water pollution remain as worsening global problems. Breathing the air in Bombay is equivalent to smoking ten cigarettes a day. Ozone depletion, now 4–5 per cent in some parts of the world, may be occurring at twice the rates previously thought. In parts of Europe, half the river water is too polluted even for industrial uses (Brown et al. 1992). *An inevitable consequence of increased flows of energy and materials in current technosystems is environmental degradation.*

In marked contrast to the finely balanced, feed-back controlled biosphere, the additive result of our technosystem-driven lives is a modified ecosystem—Ayres (1989, pp. 23–49) calls this a 'synthesphere'—in which:

- heat outputs exceed inputs and thermal balance has been lost;
- the energy base on which the synthesphere currently relies, the fossil fuels, is non-renewable and polluting;
- the cycles of natural materials such as water and carbon have become unbalanced;
- many of these materials, such as phosphorus, no longer cycle effectively (remember that they are also of finite size), but are dispersed into the environment as long-lived waste residuals which themselves further affect the environment's ability to sustain itself; and
- novel chemicals, like plastics and CFCs (chlorofluorocarbons), have been injected into the system. Other chemicals like mercury, lead and chromium have been mobilised into food chains. All of these lack natural recycling pathways and become waste residuals, often of a dangerous and polluting nature.

Unlike ecosystems, where 'value' is synonymous with maintenance of energy flows—and chemical and nutrient cycles—for life-support, in technosystems only economic 'value' is relevant. Outputs, such as heat and other waste residuals, and 'common' inputs, such as air and soil (and sometimes water), have no economic value and become 'invisible' to technosystems (Folke & Kaberger 1991). They are considered to be 'externalities', not incorporated into the cost-benefit analysis of the system. A consequence of this economic invisibility is overuse. One of the biggest tasks confronting societies is to rethink our economic systems so that they reflect the true value of environmental services and resources. Our current economic system, with its cycles and exchanges of money and labour only, its one-way flows of matter, and its exclusion of environmental 'externalities', is incompatible with the development of environmentally sustainable technologies (Daly & Cobb 1989).

Environmentally sustainable technosystems

By using the ecosystem as a model of a sustainable system, it is easy to demonstrate the unsustainability of our current technosystems. The ecosystem model also offers a guide to the kinds of changes required for technologies of the future. Any transition to sustainable technosystems must involve:

- a reduction in the mass flows of energy and resources;
- a reduction in the use of non-renewable sources of energy and increased reliance on renewable and solar energy sources;
- reduced rates of consumption;
- reduction in the energy and material content of each unit of goods and services produced;

- the minimisation of emissions and waste so that they do not exceed the assimilative capacity of the environment;
- mass recycling of materials; and
- a reduction in the rate of human population growth to a level which does not exceed the carrying capacity of the biosphere.

The task of initiating a transition to sustainable technologies is a much more difficult one. Firstly, as shown by the continuum of perspectives on sustainability, there is disagreement about what sustainability actually means. Yet some agreement seems essential for any future planning. Secondly, achieving a sustainable society requires much more than an understanding of model ecosystems operating in nature. It requires substantial behavioural and structural change at the individual, sectoral, national and international levels, and within socio-economic, cultural and political structures. It requires a global web of change set within a common sustainability perspective. But which perspective?

Analysing the four perspectives introduced earlier, within the context of our ecological framework, it is clear that each of them has deficiencies as a framework for transition to a sustainable society. All three anthropocentric perspectives remain wedded to economic growth in the traditional sense—and we define economic growth by the capacity of technology to produce more goods and services, by using more energy and resources.

The 'technofix' perspective fails to recognise that all technologies use energy and matter and rely on environmental support services. While it may make economic sense in the short term to use innovative technologies to sequentially exploit the environment, it makes no sense (economic or otherwise) in the long term. Such a strategy will always be accompanied by environmental degradation, by resource depletion and hence by a reduction in the capacity of the environmental system to support the technology itself. It is a self-defeating perspective.

Similarly, there are serious contradictions in the 'resource development' perspective, from the viewpoint of true sustainability, with its inadequate integration of economic and ecological principles. This perspective fails to bridge the gap between the economist's and the ecologist's view of a 'natural resource'.

The biocentric view of the 'extreme greenie' is a useful educative perspective in that it reminds us that humans are biological organisms, dependent on the integrity of ecological processes. However, a technological backtrack for over-consuming industrial nations is not the answer. Nor will it help address the enormous inequities facing non-industrialised nations.

The 'resource management for development' perspective is useful in recognising that some degree of global change is required to achieve a transition to sustainability. Proponents of this perspective argue that continued economic growth is essential if underdeveloped nations are to have any chance of achieving improvements in well-being. They promote sustainability, continued economic growth and equity as part of the same package. But how are we to do this? Is economic growth, as currently defined, compatible with sustainability?

The answer would seem to be 'no', and the consequent contradictions inherent in each of the the growth perspectives have given rise to radical but more logical approaches to sustainability. A new perspective has developed which, while somewhat anthropocentric, is primarily biocentric in emphasis. This new perspective is called 'eco-development' and it has spawned new approaches to technology such as eco-technology or ecological engineering (Mitsch & Jorgensen 1989).

Eco-development recognises that while some forms of economic development are necessary, there is a need to break the nexus between development and growth. This will require a qualitative redefinition of economic growth away from increased flows of energy and resources through technosystems towards systems which ensure the maintenance—or improvement—of environmental capital (Daly 1990). Eco-development embeds within it the important concepts of technology assessment and deliberate choice for ecological fitness. Integrated agro-industrial ecosystems provide excellent examples of the way forward. These ecosystems employ environmentally responsible methods for the production of food, energy and chemicals (Tiezzi et al. 1991). This is an essential shift from the reigning viewpoint where economic fitness is the primary, and often the only, requirement for technological development. How might we move forward from such a perspective?

Directions towards sustainability

The transition to sustainable societies will require a rethinking of the way in which we assess, choose and use technology. It will also need to address the issues of inequity of wealth and resource use, and of overconsumption and overpopulation.

At the individual level, we will need to redirect our consumption patterns towards those which reflect and support sustainability of energy and resource use. Local communities will need to play ever-increasing roles in recycling and other resource-conserving practices. But real change cannot occur without major restructuring within the national agencies where budgetary power and economic decision-making reside. A free-market-based economic democracy cannot evolve by itself towards a sustainable system. It will require policies and expenditures which reflect and encourage sustainable values in the market place. This includes incentives for technological innovation which replace current unsustainable technologies with those which are conserving of energy and resources, orientated towards renewable energy sources. Such technologies must be integrated with natural recycling pathways to minimise waste production. There is no shortage of target areas for change and development at the national level:

- the removal of public policies which encourage environmentally unsustainable practices;
- restructuring of tax systems to provide appropriate incentives (for material recycling, resource conservation, technological innovation for sustainability) and disincentives (for polluting and resource-wasteful technologies, for overconsumption of goods and services);
- specific policies to encourage the development of technologies which use renewable energy sources; and
- the reform of the economic system so that it is fully assimilated into an ecological framework, and able to internalise the environmental costs associated with technology use and development.

Many areas for change require cooperation on an international scale. Foreign policy—for example, that dealing with trade—will need to reflect the sustainable goals of national policies. Perhaps the most difficult barrier to sustainability is the inequitable distribution of wealth and resource use. People without the ability to sustain themselves do not have the luxury of debating the issue of global sustainability. Foreign policy, particularly that of industrialised nations, will need to recognise an urgent need for equity, ensuring that aid programs are compatible with sustainable goals and seriously penalising resource exploitation by wealthy nations. It will be a difficult task to conserve and rehabilitate what we have left without further economic marginalisation of underdeveloped nations.

Conclusion

Change needs to occur locally, nationally and globally. Local communities, by their actions and behaviours, provide an important stimulus for institutional change. Yet only national and international institutions have the regulatory, economic and budgetary power to provide the appropriate incentives for change on a nation-state and regional level. There are indications, however, that environmentally responsible changes have begun to take place nationally and internationally, albeit progressing slowly (Starke 1990). In addition to formal institutions, independent environmental groups such as Greenpeace, operating locally and globally, are catalysts for change. They bridge the gap between community concerns and national responses.

Technocrats among us argue that radical directional change towards environmental sustainability will cause major social and economic dislocation. The ecologists argue that we have no choice. If we choose to do nothing the environment and the laws of thermodynamics will ultimately constrain us—as they are beginning to do—resulting in far greater upheaval. Better then, to plan and develop the necessary structures for an orderly transition to sustainability which minimises social and political crisis. What do you think?

References

Ayres, Robert U. 1989, 'Industrial metabolism', *Technology and environment*, eds Jesse H. Ausubel & Hedy E. Sladovich, National Academy Press, Washington DC, pp. 23–49

Barr, Neil and Cary, John 1992, *Greening a brown land: the Australian search for sustainable land use*, Macmillan Education, Australia

Brown, Lester 1991, *States of the world 1991*, Worldwatch Institute, W. W. Norton & Co., New York

Brown, Lester, Flavin, Christopher and Postel, Sandra 1992, *Saving the planet: how to shape an environmentally sustainable economy*, Earthscan Publications Ltd., London

Daly, Herman E. 1990, 'Toward some operational principles of sustainable development', *Ecological economics*, no. 2, pp. 1–6

Daly, Herman E. and Cobb, John B. 1989, *For the common good: redirecting the economy toward community, the environment, and a sustainable future*, Beacon Press, Boston, ch. 4, pp. 85–96

Folke, Carl 1991, 'Socio-economic dependence on the life-supporting environment', *Linking the natural environment and the economy: essays from the Eco-Eco-Group*, eds Carl Folke & Tomas Kaberger, Kluwer Academic Publishers, London, ch. 5

Folke, Carl and Kaberger, Tomas 1991, 'Recent trends in linking the natural environment and the economy', *Linking the natural environment and the economy: essays from the Eco-Eco-Group*, eds Carl Folke & Tomas Kaberger, Kluwer Academic Publishers, London, pp. 273–300

Gifford, R. M. 1976, 'An overview of fuel used for crops and national agricultural systems', *Search*, vol. 7, pp. 412–17

Mitsch, William and Jorgensen, Sven Erik, 1989, *Ecological engineering: an introduction to ecotechnology*, John Wiley & Sons, New York

Rappoport, R. A. 1971, 'The flow of energy in an agricultural society', *Energy and power*, Scientific American, US

Sabath, Michael D. and Quinell, Susan 1981, *Ecosystems, energy and materials: the Australian context*, Longman Cheshire, Australia

Starke Linda 1990, *Signs of hope: working towards our common future*, Oxford University Press, Oxford

Tiezzi, E., Marchettini, N. and Ulgiati, S. 1991, 'Integrated agro-industrial ecosystems: an assessment of the sustainability of a cogenerative approach to food, energy and chemicals production by photosynthesis', *Ecological economics: the science and management of sustainability*, ed. Robert Constanza, Columbia University Press, New York, ch. 30, pp. 459–73

WCED (World Commission for Environment and Development), 1987, *Our common future*, Oxford University Press, Oxford

Annotated bibliography

Beckerman, W. 1977, 'The fallacy of finite resouces', *Solar Australia: Australia at the crossroads*, Ambassador Press, NSW, Appendix C, pp. 116–21
A rare expression of the technofix philosophy.

Daly, Herman E. and Cobb, John B. 1989, *For the common good; redirecting the economy toward community, the environment, and a sustainable future*, Beacon Press, Boston, ch. 4, pp. 85–96
Economic models paint humanity with insatiable wants and as totally individualistic. Can they be changed?

Folke, Carl 1991, 'The societal value of wetland support', *Linking the natural environment and the economy: essays from the Eco-Eco-Group*, eds Carl Folke & Tomas Kaberger, Kluwer Academic Publishers, London, ch. 8, pp. 141–71
Folke demonstrates that when we remove support services such as wetlands, the cost of replacing them with technology can be millions of dollars per year.

Gray, Paul 1989, 'The paradox of technological development', *Technology and environment*, National Academy Press, Washington, D.C., pp. 192–204
Gray offers a positive perspective on the need for continuing developing nuclear power technology. His writing is a mix of technofix/environmentalist arguments.

Harvey, David 1974, 'Population, resources, and the ideology of science', *Economic geography*, no. 50, pp. 256–77
Compare this thoughtful article with that of Roger Short 1991,'Human population growth', *Habitat Australia*, October, pp. 13–14. Do we slip into an elitist ideology when debating the human population problem?

Slatyer, R.O. 1972, 'Energy flows in the biosphere—the impact of man', *Proceedings of the Linnean Society of New South Wales*, vol. 97, pt 3, pp. 227–36

In 1972 an Australian scientist warned of global warming and asked if we were prepared to change our lifestyles of endless consumption and planned obsolescence.

CREDITS

The editors wish to thank the following for permission to reproduce copyright material.

(p. 7) David Noble, *Forces of production*, copyright 1984 by David F. Noble, by permission of Random House; (p. 48) Wendy Harmer, 'Gadget, greed's ghosts', *21.C*, by permission of the author; (pp. 54–6) Ian Miles, *Social indicators for human development*, by permission of Pinter Publishers, all rights reserved; (pp. 60, 65–6) B. Koenig, 1988, 'The technological imperative in routine medical practice: the social creation of a "routine" treatment', *Biomedicine examined*, eds M. Lock & D. Gordon, Kluwer Academic Publishers, Dordrecht, pp. 465–96, copyright 1988 by Kluwer Academic Publishers, reprinted by permission of Kluwer Academic Publishers; (p. 79) Cees J. Hamelink, 1988, *The technology gamble. Informatics and public policy: a study of technology choice*, Ablex, Norwood N. J., copyright 1988 by Ablex Publishing Corporation, by permission of Ablex Publishing Corporation, all rights reserved; (p. 80) N. J. Vig, 'Technology, philosophy and the State' in *Technology and politics* (Kraft & Vig, eds.) 1988, Durham, N. C., Duke University Press. Reprinted by permission of the publisher; (pp. 83, 84, 85) K. Dyson and P. Humphreys, *The political economy of communications: international and European dimensions*, by permission of Routledge; (p. 86) James G. Savage, *The politics of international telecommunications regulation*, by permission of the author; (pp. 105, 108, 109, 113, 114) John Keane, *The media and democracy*, by permission of Blackwell Publishers; (p. 107) C. B. Macpherson, *The life and times of liberal democracy*, 1977, copyright C. B. Macpherson, by permission of the Oxford University Press; (p. 165, 166) Frederic Jameson, 'Postmodernism, or the cultural logic of late capitalism', *New left review*, 146, pp. 53–92, by permission of the *New left review*; (p. 197) Figure adapted from Robert Ayres, 'Industrial metabolism,' *Technology and environment*, eds Jesse H. Ausubel & Hedy E. Sladovich, copyright 1989 by the National Academy of Sciences, by permission of the National Academy Press, Washington DC.

Every effort has been made to contact copyright holders. However, where an omission has occurred, the editors and publisher will gladly include acknowledgement in any future edition.

INDEX

References to the glossary have not been included. Names of persons are included only if their work is discussed in the text, while organisations mentioned or quoted in the text have been included.